# RING THEORY II

# PURE AND APPLIED MATHEMATICS

*A Program of Monographs, Textbooks, and Lecture Notes*

# LECTURE NOTES
# IN PURE AND APPLIED MATHEMATICS

*Other volumes in preparation*

# RING THEORY II

## Proceedings of the Second Oklahoma Conference

EDITED BY

### *Bernard R. McDonald and Robert A. Morris*

DEPARTMENT OF MATHEMATICS
UNIVERSITY OF OKLAHOMA
NORMAN, OKLAHOMA

MARCEL DEKKER, INC. New York and Basel

**Library of Congress Cataloging in Publication Data**

Ring Theory Conference, 2d, University of Oklahoma,
    1975.
    Ring theory II.

    (Lecture notes in pure and applied mathematics ; 26)
    Includes bibliographies.
    1.  Rings (Algebra)--Congresses.  I.  McDonald,
Bernard R.  II.  Morris, Robert A.  III.  Title.
QA247.R56  1975        512'.4        76-55134
ISBN 0-8247-6575-3

MARCEL DEKKER, INC.

270 Madison Avenue, New York, New York  10016


Current printing (last digit):
10 9 8 7 6 5 4 3 2 1


PRINTED IN THE UNITED STATES OF AMERICA

PREFACE

This volume contains a collection of the papers presented at the Ring Theory Conference held at The University of Oklahoma during March 11-13, 1975.

This conference was conceived as primarily an educational affair, its goal being to inform mathematicians about current trends in ring theory. Because of this we purposely sought a cross section of topics from both commutative and noncommutative ring theory and encouraged expository presentations. Further, we hoped this conference would complement the ring theory conference of 1973 held at The University of Oklahoma and that this monograph would provide a sequel to the proceedings of the 1973 conference. The proceedings of the 1973 conference appear as Ring Theory, Vol. 7, Lecture Notes in Pure and Applied Mathematics, Dekker (1974) (Edited by B. McDonald, A. Magid and K. Smith).

The three principal speakers were Maurice Auslander, Moss Swee3dler and Daniel Zelinsky. Professor Swee3dler noted there is a "silent 3" which recently appeared in the spelling of his name. Each of the principal speakers presented two one-hour lectures. Ten shorter survey talks were also given. All of the manuscripts of the above talks are included in this monograph and the topics and speakers are listed in the Contents.

Short research talks were given by J. Brewer, "Coherence and weak dimen-
sion of R[[X]] "; F. DeMeyer, "The Brauer group of a curve"; J.
Fisher, "Finite groups of automorphisms acting on rings"; G. Garfinkel,
"Generic N-splitting algebras"; W. Leavitt, "The upper radical"; S.
Wiegand, "Prime ideal structure in Noetherian rings"; and M. Teply,
"On the transfer of properties to subidealizer rings."

We are indebted to the authors of the papers in this monograph for
their prompt preparation and cooperation and, generally, to the partici-
pants of the conference for their interest and enthusiasm. We take
the responsibility for any misprints and errors this volume contains.

The funding for the conference was from several sources at The University
of Oklahoma. Contributing financially to the conference were Dean
Gordon Atkinson of the Graduate College, Dean Paige Mulhollan of the
College of Arts and Sciences and the Department of Mathematics whose
chairman is Gene Levy. A special note of appreciation is due Paige
Mulhollan, who opened the conference, contributed generously to the
support of its activities, and has been enthusiastic in his encourage-
ment of our algebra program.

Appreciation is also due the secretarial staff — Trish Abolins, Sonya
Fallgatter, Debbie Franke, Rose Kendrick, Denise Kueny — and the
faculty and graduate students of our department whose efforts and
contributions made the conference a success.  We were fortunate to
have an excellent typist, Trish Abolins, who is patient, efficient and
helpful.


B. R. McDonald

R. A. Morris

# LIST OF CONTRIBUTORS

M. AUSLANDER, Brandeis University, Waltham, Massachusetts

R. K. DENNIS, Cornell University, Ithaca, New York

E. GREEN, University of Illinois, Urbana, Illinois

K. GOODEARL, University of Utah, Salt Lake City, Utah

M. HOCHSTER, Purdue University, West Lafayette, Indiana

L. LEVY, University of Wisconsin, Madison, Wisconsin

M. SCHACHER, University of California, Los Angeles, California

M. SMITH, University of Texas, Austin, Texas

R. SNIDER, Virginia Polytechnic Institute and State University, Blacksburg,
    Virginia

M. SWEE3DLER, Cornell University, Ithaca, New York

R. WIEGAND, University of Nebraska, Lincoln, Nebraska

D. ZELINSKY, Northwestern University, Evanston, Illinois

J. ZELMANOWITZ, University of California, Santa Barbara, California

# LIST OF PARTICIPANTS

| | |
|---|---|
| Armendariz, Efraim P. | University of Texas |
| Auslander, Maurice | Brandeis University |
| Bahl, Christina | Drake University |
| Beachy, John | Northern Illinois University |
| Bertholf, Dennis | Oklahoma State University |
| Blair, William | Northern Illinois University |
| Boyle, Ann | University of Wisconsin, Milwaukee |
| Brewer, Jim | University of Kansas |
| Burgess, W. D. | University of Ottawa |
| Crown, Gary | Wichita State University |
| DeMeyer, Frank | Colorado State University |
| Dennis, R. Keith | Cornell University |
| Diamond, Bob | University of Oklahoma |
| Dickson, Spencer | Iowa State University |
| Eliason, Stanley | University of Oklahoma |
| Fenrick, Maureen | Wichita State University |
| Fieldhouse, David | University of Guelph |
| Fisher, Joe W. | University of Cincinnati |
| Fuelberth, John | University of Northern Colorado |
| Fuller, Kent | University of Iowa |
| Garfinkel, Gerald | New Mexico State University |
| Goodearl, Ken | University of Utah |

Gordon, Robert                          Temple University

Green, Edward L.                        University of Illinois

Gustafson, William                      Indiana University

Hausknecht, Adam O.                     University of Kansas

Heinicke, Allan                         University of Western Ontario

Henderson, Sydney                       University of Oklahoma

Herrington, Mary                        University of Oklahoma

Hershberger, Bessie                     University of Oklahoma

Hochster, Melvin                        Purdue University

Kao, Shaw-Yi                            University of Oklahoma

Kirkman, Ellen E.                       Wake Forest University

Koether, Robb T.                        University of Oklahoma

Krause, G.                              University of Manitoba

Kuzmanovich, Jim                        Wake Forest University

Lady, Everett Lee                       University of Kansas

Leavitt, W. G.                          University of Nebraska

Levy, Gene                              University of Oklahoma

Levy, Lawrence                          University of Wisconsin

Lewin, Jacques                          Syracuse University

Lewis, William James                    University of Nebraska

Macrae, R. E.                           University of Colorado

Magid, Andy                             University of Oklahoma

Matchett, Andrew J.                     Texas A & M

McAdam, Steve                           University of Texas

McDonald, Bernie                        University of Oklahoma

| | |
|---|---|
| Moore, Jeanna | University of Oklahoma |
| Morris, Robert A. | University of Oklahoma |
| Nakao, Zensho | University of Oklahoma |
| Naylor, Larry | University of Oklahoma |
| Oldroyd, Andrew | University of Oklahoma |
| Olson, Dwight | Cameron University |
| Osterburg, James | University of Cincinnati |
| Pittman, Gary | University of Oklahoma |
| Roark, Chuck | University of Oklahoma |
| Rossa, Robert | Arkansas State University |
| Rowe, David | University of Oklahoma |
| Rubin, Leonard | University of Oklahoma |
| Rutter, Ed | University of Kansas |
| St. Mary, Donald | University of Massachusetts |
| Schacher, Murray | U.C.L.A. |
| Shapiro, Jay | Wayne State University |
| Shirley, Frank | University of Texas |
| Shores, Thomas | University of Nebraska |
| Smith, Kirby | Texas A & M |
| Smith, Martha K. | University of Texas |
| Snider, Robert L. | V.P.I. & S.U. |
| Steury, Lawrence | University of Oklahoma |
| Svec, Michael | University of Oklahoma |
| Swee3dler, Moss | Cornell University |
| Tan, Richard | University of Oklahoma |

Tangeman, Dick                          Arkansas State University

Teply, Mark L.                          University of Florida

Walker, Elbert A.                       New Mexico State University

Walls, Gary Lee                         Oklahoma State University

Walsh, John                             University of Oklahoma

Wang, Stuart                            University of Oklahoma

Wiegand, Roger                          University of Nebraska

Wiegand, Sylvia                         University of Nebraska

Woods, Sheila M.                        University of Manitoba

Zelinsky, Daniel                        Northwestern University

Zelmanowitz, Julius                     U.C.S.B.

# CONTENTS

## POWER-CANCELLATION OF MODULES                                      131

### K. R. Goodearl

## DIAGRAMMATIC TECHNIQUES IN THE STUDY OF INDECOMPOSABLE MODULES      149

### Edward L. Green

# RING THEORY II

EXISTENCE THEOREMS FOR ALMOST SPLIT SEQUENCES

Maurice Auslander[1]

Brandeis University

§1. **Introduction.** In studying the representation theory of Artin
algebras, which includes finite dimensional algebras over a field as a
special case, I. Reiten and I introduced in [3], [4] the notion of an
almost split sequence. This has proven to be of considerable use in
studying modules over Artin algebras (see [1], [3], [4], [5]).

Naturally it is reasonable to ask if this notion is of significance
only for Artin algebras or if it can be extended to more general types
of rings. In these lectures we will discuss the situation for
Noetherian algebras over complete local Noetherian rings, indicating
the basic existence theorems as well as giving a sampling of applications. The list of applications is not meant to be exhaustive, just
indicative of the wide variety of questions to which these ideas are applicable.

For the most part proofs have been omitted. The interested reader
is referred to the bibliography for proofs analogous to those in the

[1] Written with partial support of NSF Grant MPS 997-444-0276.

literature.  A more complete exposition of this material will appear in the future.

Finally, I would like to take this opportunity to express my appreciation for being invited to address The Oklahoma University Ring Theory Conference.

§2.  **Basic Existence Theorem**. By an **R-algebra** $\Lambda$ we mean a commutative ring $R$ together with a ring morphism $f : R \to \Lambda$ with $f(R)$ contained in the center of $\Lambda$ . We say that an R-algebra $\Lambda$ is a **Noetherian R-algebra** if $R$ is a Noetherian ring and $\Lambda$ is finitely generated as an R-module.  Clearly if $\Lambda$ is a Noetherian R-algebra, then $\Lambda$ is both a left and right Noetherian ring.  We say that an R-algebra $\Lambda$ is an **Artin R-algebra** if $R$ is an Artin ring and $\Lambda$ is finitely generated as an R-module.  Clearly if $\Lambda$ is an Artin R-algebra, then $\Lambda$ is both a left and right Artinian ring.  Also each Artin R-algebra is a Noetherian R-algebra.  Since our primary interest in these lectures is Noetherian R-algebras with $R$ a complete local ring, we will assume from now on, unless stated to the contrary, that all R-algebras are Noetherian R-algebras with $R$ a complete local ring.  We now turn our attention to the notion of almost split sequences.  The reader is referred to [4] and [5] for the basic definition and proofs concerning almost split sequences.

Let $\Gamma$ be an arbitrary ring and $\text{Mod}(\Gamma)$ the category of left $\Gamma$-modules.  We view right $\Gamma$-modules as left $\Gamma^{op}$-modules where $\Gamma^{op}$ is the **opposite ring** of $\Gamma$ .  So from now on a $\Gamma$-module always means a left

Γ-module.  An exact sequence  (*)  $0 \longrightarrow A \xrightarrow{u} B \xrightarrow{v} C \longrightarrow 0$  in

Mod(Γ)  is said to be an **almost split sequence** if it satisfies the fol-

lowing conditions:

   (a)  it is not a splittable sequence;

   (b)  if  $g : X \to C$  is <u>not</u> a splittable epimorphism (a morphism
        $g : X \to C$  is said to be a **splittable epimorphism** if there is
        a morphism  $t : C \to X$  such that  $gt = id_C$ ), then there is a
        morphism  $j : X \to B$  such that  $vj = g$ .

   (c)  if  $h : A \to Y$  is <u>not</u> a splittable monomorphism (a morphism
        $h : A \to Y$  is said to be a **splittable monomorphism** if there is
        a morphism  $s : Y \to A$  such that  $sh = id_A$ ), then there is a
        morphism  $p : B \to Y$  such that  $pu = h$ .

   The following properties of almost split sequences are not difficult
to verify (see [4], [5]).


<u>Proposition 2.1.</u>  Suppose  $0 \longrightarrow A \xrightarrow{u} B \xrightarrow{v} C \longrightarrow 0$  and  $0 \longrightarrow$
$A' \xrightarrow{u'} B' \xrightarrow{v'} C' \longrightarrow 0$  are almost split sequences of Γ-modules.

   (a)  $End_\Gamma(A)$ , $End_\Gamma(C)$  are local rings (have a unique maximal left
        ideal) and so  A  and  C  are indecomposable Λ-modules.

   (b)  The following statements are equivalent:

        (i)  The sequences  $0 \longrightarrow A \xrightarrow{u} B \xrightarrow{v} C \longrightarrow 0$  and

             $0 \longrightarrow A' \xrightarrow{u'} B' \xrightarrow{v'} C' \longrightarrow 0$  are isomorphic

             i.e., there is a commutative diagram

$$
\begin{array}{ccccccccc}
0 & \longrightarrow & A & \xrightarrow{u} & B & \xrightarrow{v} & C & \longrightarrow & 0 \\
  &   & \downarrow &   & \downarrow &   & \downarrow &   &   \\
0 & \longrightarrow & A' & \xrightarrow{u'} & B' & \xrightarrow{v'} & C' & \longrightarrow & 0
\end{array}
$$

with the vertical morphisms isomorphisms,

(ii)   $A \approx A'$ ,

(iii)   $C \approx C'$ .

Thus we see that an almost split sequence $0 \to A \to B \to C \to 0$ is determined uniquely (up to isomorphism) by either $A$ or $C$ and so is an invariant of $C$ and of $A$ when it exists. These observations suggest the following questions.

(1)  For which $\Gamma$-modules $A$ are there almost split sequences $0 \to A \to B \to C \to 0$ ?

(2)  For which $\Gamma$-modules $C$ are there almost split sequences $0 \to A \to B \to C \to 0$ ?

(3)  Suppose $A$ and $C$ are connected by an almost split sequence $0 \to A \to B \to C \to 0$ . Then $A$ uniquely determines $C$ and $C$ uniquely determines $A$ . Give as explicit a way as possible of describing $A$ in terms of $C$ and $C$ in terms of $A$ .

(4)  What information concerning $C$ or $A$ does one need to be able to construct its uniquely determined almost split sequence $0 \to A \to B \to C \to 0$ , assuming, of course, that it exists?

(5)  What do the properties of the almost split sequence $0 \to A \to B \to C \to 0$ tell one about $A$ or about $C$ and the other way around?

While even for arbitrary Artin R-algebras, let alone arbitrary rings $\Gamma$ , no complete answers exist for these questions, a considerable amount can be said concerning some of these for R-algebras with $R$ a complete local ring. So we now turn our attention to this situation.

Suppose $\Lambda$ is an R-algebra (with R a complete local ring). We can now state our main existence theorem for almost split sequences in $Mod(\Lambda)$ in a preliminary form.

<u>Theorem 2.2.</u>  (a)  If C is an indecomposable nonprojective module in
Noeth($\Lambda$) (where Noeth($\Lambda$) denotes the category of Noetherian $\Lambda$-modules) then there is an almost split sequence $0 \to A \to B \to C \to 0$ . This sequence has the property that A is an indecomposable noninjective $\Lambda$-module in Art($\Lambda$) (where Art($\Lambda$) denotes the category of Artin $\Lambda$-modules).

(b)  If A is an indecomposable noninjective module in Art($\Lambda$) , then there is an almost split sequence $0 \to A \to B \to C \to 0$ . This sequence has the property that C is an indecomposable nonprojective module in Noeth($\Lambda$) .

This theorem was proven in [4] for the special case where R is Artinian rather than just a complete Noetherian local ring. Of course, $\Lambda$ is Artinian when R is Artinian, so Art($\Lambda$) = Noeth($\Lambda$) , in this case. But except for this change the proof for R Artinian can be copied verbatim for R a complete local ring. In order to give the definitive statement of the existence theorem for almost split sequences as well as indicate how to go from the Artin case to the complete local Noetherian case, we point out some of the important properties of algebras over complete local rings used in the statement and proof of theorem.

Let $\Lambda$ be an R-algebra where R is a complete local ring with maximal ideal $\underline{m}$ . Let $I = I(R/\underline{m})$ be an injective envelope of $R/\underline{m}$ over $R$ . Define the functor $D : Mod(\Lambda) \rightarrow Mod(\Lambda^{op})$ by $D(X) =$ $Hom_R(X,I)$ for all $X$ in $Mod(\Lambda)$ . Define $D : Mod(\Lambda^{op}) \rightarrow Mod(\Lambda)$ similarly. For each $X$ in $Mod(\Lambda)$ define $\phi_X : X \rightarrow Hom_R(Hom_R(X,I),I)$ by $\phi_X(x)(f) = f(x)$ for all $x$ in $X$ and $f$ in $Hom_R(X,I)$ . Since $R$ is complete it is well-known by Matlis duality that $X$ is in $Noeth(\Lambda)$ ( $Noeth(\Lambda^{op})$ ) if and only if $D(X)$ is an $Art(\Lambda^{op})$ ( $Art(\Lambda$ Thus the functors $D$ induce functors $D : Noeth(\Lambda) \rightarrow Art(\Lambda^{op})$ and $Art(\Lambda^{op}) \rightarrow Noeth(\Lambda)$ with the property that $\phi_X : X \rightarrow D^2(X)$ is an isomorphism for $X$ in $Noeth(\Lambda)$ or in $Art(\Lambda^{op})$ . Thus we have a duality $D : Noeth(\Lambda) \rightarrow Art(\Lambda^{op})$ .

Since $R$ is complete any R-algebra $\Gamma$ is semiperfect (i.e., every finitely generated $\Gamma$-module has a projective cover or, equivalently, $\Gamma/rad(\Gamma)$ is semisimple and every idempotent in $\Gamma/rad(\Gamma)$ is the image of an idempotent in $\Gamma$ ). Therefore since for each $X$ in $Noeth(\Lambda)$ , we know that $End_\Lambda(X)$ is an R-algebra, it follows that $End_\Lambda(X)$ is semiperfect for each $X$ in $Noeth(\Lambda)$ . Hence $End_\Lambda(Y)$ is an R-algebra and therefore semiperfect for all $Y$ in $Art(\Lambda)$ , since $D(Y)$ is in $Noeth(\Lambda^{op})$ and $End_\Lambda(Y)^{op} \cong End_{\Lambda^{op}}(D(Y))$ by the duality $D : Noeth(\Lambda)$ $Art(\Lambda^{op})$ . Therefore $X$ in $Noeth(\Lambda)$ or in $Art(\Lambda)$ is indecomposable if and only if $End_\Lambda(X)$ is local and so by the usual Krull-Schmidt arguments every $X$ in $Noeth(\Lambda)$ or $Art(\Lambda)$ can be written uniquely (up to isomorphism) as a finite sum (direct) $X_1 \amalg \cdots \amalg X_n$ with the $X_i$ indecomposable X-modules. That is why in the statement of Theorem 2.2 we had to only assume that $C$ and $A$ are indecomposable since this

is equivalent to the statement that $\text{End}_\Lambda(C)$ and $\text{End}_\Lambda(A)$ are local,
a necessary condition for an almost split sequence $0 \to A \to B \to C \to 0$
to exist.

Next we recall the definition of the transpose. Let $\text{Noeth}_P(\Lambda)$
denote the full subcategory of $\text{Noeth}(\Lambda)$ whose objects are the $X$ in
$\text{Noeth}(\Lambda)$ which have no nonzero projective summands. Associated with
$\text{Noeth}_P(\Lambda)$ is the category $\underline{\text{Noeth}}_P(\Lambda)$ whose objects are the same as those
in $\text{Noeth}_P(\Lambda)$ and where $\underline{\text{Hom}}(X,Y)$ in $\underline{\text{Noeth}}_P(\Lambda)$ is defined to be
$\text{Hom}_\Lambda(X,Y)/P(X,Y)$ where $P(X,Y)$ is the R-submodule of $\text{Hom}_\Lambda(X,Y)$
consisting of all morphisms $f : X \to Y$ such that there is a factoriza-
tion $X \to P \to Y$ with $P$ projective. Then $\underline{\text{Noeth}}_P(\Lambda)$ is an additive
category. There is a duality $\text{Tr} : \underline{\text{Noeth}}_P(\Lambda) \to \underline{\text{Noeth}}_P(\Lambda^{op})$ called the
**transpose** which is defined as follows.

For each $X$ in $\text{Noeth}_P(\Lambda)$ , let $P_1 \to P_0 \to X \to 0$ be a **minimal**
**projective** **presentation** of $X$ , i.e., $P_0 \to X$ is a projective cover
and $P_1 \to \ker(P_0 \to X)$ is a projective cover. Then define $\text{Tr}(X) =$
$\text{coker}(P_0^* \to P_1^*)$ where $Z^*$ is the $\Lambda^{op}$-module $\text{Hom}_\Lambda(Z,\Lambda)$ in $\text{Noeth}_P(\Lambda^{op})$ .
While $\text{Tr}(X)$ is well-defined on objects in $\text{Noeth}_P(\Lambda)$ it is not well-
defined on morphisms $X \to Y$ since there is no unique way to "lift" a
morphism $X \to Y$ to a morphism from a minimal projective presentation
of $X$ to that of $Y$ . However, it does give a well-defined contra-
variant functor $\text{Tr} : \underline{\text{Noeth}}_P(\Lambda) \to \underline{\text{Noeth}}_P(\Lambda^{op})$ called the transpose.
Since $\text{Tr} : \underline{\text{End}}_\Lambda(X) \cong \underline{\text{End}}_{\Lambda^{op}}(\text{Tr}(X))^{op}$ , it follows that $X$ is indecomposable if and only if $\text{Tr}(X)$ is indecomposable. This is because for

Z  in  $\text{Noeth}_p(\Lambda)$ ,  $\text{End}_\Lambda(Z)$  is local if and only if  $\underline{\text{End}}_\Lambda(Z)$  is local.

Now combining the dualities  $D : \text{Noeth}(\Lambda) \to \text{Art}(\Lambda^{op})$  and  $\text{Tr} :$

$\underline{\text{Noeth}}_p(\Lambda) \to \underline{\text{Noeth}}_p(\Lambda^{op})$  we obtain an equivalence  $D\text{Tr} : \underline{\text{Noeth}}_p(\Lambda) \to$

$\overline{\text{Art}}_I(\Lambda)$  where  $\overline{\text{Art}}_I(\Lambda)$  is the subcategory of  $\text{Art}_I(\Lambda)$  consisting of a

X  in  $\text{Art}(\Lambda)$  with no nonzero injective summands and  $\overline{\text{Art}}_I(\Lambda)$  has the

same objects as  $\text{Art}_I(\Lambda)$  with morphisms  $\overline{\text{Hom}}_\Lambda(X,Y) = \text{Hom}(X,Y)/I(X,Y)$

where  $I(X,Y)$  consists of all  $f : X \to Y$  which can be factored as

$X \to I \to Y$ with  I  injective.  Then it is easily seen that  $D : \text{Noeth}(\Lambda^{op})$

$\text{Art}(\Lambda)$  induces a duality  $\underline{\text{Noeth}}_p(\Lambda^{op}) \to \overline{\text{Art}}_I(\Lambda)$  so the composition

$\underline{\text{Noeth}}_p(\Lambda) \xrightarrow{\text{Tr}} \underline{\text{Noeth}}_p(\Lambda^{op}) \xrightarrow{D} \overline{\text{Art}}_I(\Lambda)$  gives an equivalence of categor

whose inverse is  $\text{Tr}(D)$ .

Since as we have observed earlier  X  in  $\text{Noeth}_p(\Lambda)$  is indecomposal

if and only if  $\text{Tr}(X)$  in  $\text{Noeth}_p(\Lambda^{op})$  is indecomposable it follows that

X  in  $\text{Noeth}(\Lambda)$  is indecomposable if and only if  $D\text{Tr}(X)$  in  $\text{Art}_I(\Lambda)$

is indecomposable.  Thus the equivalence of categories  $D\text{Tr} : \text{Noeth}_p(\Lambda)$ 

$\overline{\text{Art}}_I(\Lambda)$  gives a bijection between isomorphism classes of indecomposable

modules in  $\text{Noeth}_p(\Lambda)$  and isomorphism classes of indecomposable modules

in  $\text{Art}_I(\Lambda)$  which as we now see is precisely the correspondence given

by almost split sequences as described in Theorem 2.2.

Theorem 2.3.  Let  $\Lambda$  be an R-algebra with  R  a complete local ring.

Suppose  C  is an indecomposable module in  $\text{Noeth}_p(\Lambda)$ .  Then,

(a)  $D\text{Tr}(C)$  is an indecomposable module in  $\text{Art}_I(\Lambda)$ .

(b)  $\text{Ext}^1_\Lambda(C,D\text{Tr}(C)) \approx \text{Hom}_R(\underline{\text{End}}(C),I)$  as an  $\underline{\text{End}}(C)^{op}\text{-}\underline{\text{End}}(C)$

bimodule where  $\text{Ext}^1_\Lambda(C,D\text{Tr}(C))$  is considered a module over

$End(C)^{op}$ by means of the operation of $End(C)$ on $C$ and a module over $End(C)$ by means of the isomorphism $DTr$ : $End(C) \to \overline{End}(DTr(C))$ .

(c)  The socle of $Ext_\Lambda^1(C,DTr(C))$ is the same when viewed as an $End(C)^{op}$-module and as an $End(C)$-module and is simple in either case. Moreover, $Ext_\Lambda^1(C,DTr(C))$ is an injective envelope of its simple socle when viewed as an $End(C)^{op}$-module or as an $End(C)$-module.

(d)  The following are equivalent for an extension $0 \to DTr(C) \to E \to C \to 0$ in $Ext_\Lambda^1(C,DTr(C))$ :

   (i)  $0 \to DTr(C) \to E \to C \to 0$ is an almost split sequence,

   (ii)  $0 \to DTr(C) \to E \to C \to 0$ is a nonzero element of the socle of $Ext_\Lambda^1(C,DTr(C))$ viewed as either an $End(C)^{op}$-module or as an $End(C)$-module.

(e)  There exists an almost split sequence $0 \to DTr(C) \to E \to C \to 0$ in $Ext_\Lambda^1(C,DTr(C))$ .


We also have the companion result.


Theorem 2.4.  Let $\Lambda$ be an R-algebra with $R$ a complete local ring. Suppose $A$ is an indecomposable module in $Art_I(\Lambda)$ . Then,

(a)  $TrD(A)$ is an indecomposable module in $Noeth_p(\Lambda)$ .

(b)  $Ext_\Lambda^1(TrD(A),A) \cong Hom_R(\overline{End}(A),I)$ as an $\overline{End}(A)^{op}-\overline{End}(A)$ bimodule.

(c) $\text{Ext}_\Lambda^1(\text{TrD}(A),A)$ is an injective envelope of its simple socle
when viewed as either as an $\overline{\text{End}}(A)^{\text{op}}$ or $\overline{\text{End}}(A)$-module.
Moreover, these two socles coincide.

(d) The following are equivalent for an extension $0 \to A \to E \to$
$\text{TrD}(A) \to 0$ in $\text{Ext}_\Lambda^1(\text{TrD}(A),A)$ :

(i) $0 \to A \to E \to \text{TrD}(A) \to 0$ is an almost split sequence,

(ii) $0 \to A \to E \to \text{TrD}(A) \to 0$ is a nonzero element of the
socle of $\text{Ext}_\Lambda^1(\text{TrD}(A),A)$ viewed as either an $\overline{\text{End}}(A)^{\text{op}}$
or $\overline{\text{End}}(A)$-module.

(e) There exists an almost split sequence $0 \to A \to E \to \text{TrD}(A) \to 0$
in $\text{Ext}_\Lambda^1(\text{TrD}(A),A)$ .

As stated earlier, the proofs of Theorems 2.3 and 2.4 are essentially the same as those given in [4] for the special case $R$ is an Artin
ring.

§3. **Criteria for Almost Split Sequences.** Throughout this section we
assume that $R$ is a complete local ring and $\Lambda$ is an R-algebra. In
this section we give various criteria for when a nonsplittable sequence
$0 \to A \to B \to C \to 0$ with $A$ in $\text{Art}_I(\Lambda)$ and $C$ in $\text{Noeth}_P(\Lambda)$ is an
almost split sequence. These criteria not only facilitate finding almost
split sequences but also help to illustrate some of the properties
they possess. Some of the results given here are straightforward
generalizations of results already established for the special case

where  R  is Artin and proofs for these can be found in [4] and [5].
Others are vacuous in the case  R  is Artin.  Proofs of these will appear in a forthcoming publication.

We begin with the following problem.  Suppose  C  is an indecompos-
able module in  $\text{Noeth}_p(\Lambda)$ .  Which nonsplittable exact sequences  $0 \to$
$\text{DTr}(C) \to E \to C \to 0$  are almost split sequences?  The most elementary
question along these lines is:  are there necessarily nonsplittable
exact sequences  $0 \to \text{DTr}(C) \to E \to C \to 0$  which are not almost split
sequences?  A complete answer to this question is given in the fol-
lowing easy consequence of the fact that  $\text{Ext}_\Lambda^1(C,\text{DTr}(C))$  is an injective
envelope of its simple socle over  $\underline{\text{End}}(C)^{\text{op}}$ .

<u>Proposition 3.1.</u>  Let  C  be an indecomposable module in  $\text{Noeth}_p(\Lambda)$ .
Then the following statements are equivalent:

(a)  Every nonsplittable exact sequence  $0 \to \text{DTr}(C) \to E \to C \to 0$
is an almost split sequence.

(b)  $\underline{\text{End}}(C)$  is a division ring.

While it is possible to give examples of indecomposable  C  in
$\text{Noeth}_p(\Lambda)$  with  $\underline{\text{End}}(C)$ , a division ring, for instance  C  a simple
nonprojective module, it is more typical that  $\underline{\text{End}}(C)$  is not a division
ring.  This fact suggests it would be useful to have some special criteria
for when  $0 \to \text{DTr}(C) \to E \to C \to 0$  is an almost split sequence.  A result
along these lines is the following (see [5] for proof) proposition.

Proposition 3.2.  Let  C  be an indecomposable module in  $\text{Noeth}_p(\Lambda)$ .

The following statements are equivalent for a nonsplit exact sequence

$$0 \longrightarrow DTr(C) \xrightarrow{g} B \xrightarrow{h} C \longrightarrow 0 :$$

   (a)  $0 \longrightarrow DTr(C) \xrightarrow{g} B \xrightarrow{h} C \longrightarrow 0$  is an almost split sequence;

   (b)  for each proper submodule  C'  of  C , the exact sequence

          $0 \to DTr(C) \to h^{-1}(C') \to C' \to 0$  is splittable;

   (c)  for each nonzero submodule  A  of  DTr(C) , the exact sequence

          $0 \to DTr(C)/A \to B/A \to C \to 0$  is splittable.

     The importance of Proposition 3.2 is that it reduces the question
of whether a nonsplittable exact sequence  $0 \to DTr(C) \to B \to C \to 0$  is
an almost split sequence to a problem concerning the internal structure
of the sequence rather than one involving all possible morphisms to  C
or from  DTr(C)  as seems the case from the definition of almost split
sequences.

     While Proposition 3.1 gives some information concerning when an
exact sequence of the form  $0 \to DTr(C) \to B \to C \to 0$   (with  C  an
indecomposable module in  $\text{Noeth}_p(\Lambda)$ ), is an almost split sequence it
tells us nothing about when an exact sequence  $0 \to A \to B \to C \to 0$  is
an almost split sequence if we know only that  A  and  C  are indecom-
posable modules in  $\text{Art}_I(\Lambda)$  and  $\text{Noeth}_p(\Lambda)$ , respectively.  The rest
of this section is devoted to giving various criteria for when such
exact sequences are almost split.  These results are based on the following generally useful easy consequence of the existence of almost split
sequences.

Lemma 3.3.  Let  $0 \longrightarrow A \xrightarrow{g} B \xrightarrow{h} C \longrightarrow 0$  be a nonsplittable

exact sequence in  Mod($\Lambda$) .

(a)  If  C  is an indecomposable module in  Noeth$_p$($\Lambda$) , then there

is an exact commutative diagram

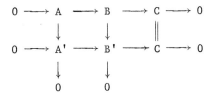

where  A'  is an indecomposable module in  Art($\Lambda$)  and the

bottom exact row is not splittable.

(b)  If  A  is an indecomposable module in  Art($\Lambda$) , there is an

indecomposable Noetherian submodule  C'  of  C  such that the

exact sequence  $0 \to A \to h^{-1}(C') \to C' \to 0$  is not splittable.

One consequence of this lemma is the following proposition.

Proposition 3.4.  Let  $0 \longrightarrow A \xrightarrow{g} B \xrightarrow{h} C \longrightarrow 0$  be a nonsplit-

table exact sequence with  A  and  C  indecomposable $\Lambda$-modules in

Art$_I$($\Lambda$)  and  Noeth$_p$($\Lambda$) , respectively.  Then the following are equivalent:

(a)  $0 \longrightarrow A \xrightarrow{g} B \xrightarrow{h} C \longrightarrow 0$  is an almost split sequence.

(b)  If  $f : H \to C$  is a morphism in  Noeth($\Lambda$)  which is not a

splittable epimorphism, then there is a morphism  $s : H \to B$

such that  hs = f .

(c)  If  $f : A \to J$  is a morphism in  Art($\Lambda$)  which is not a splittable monomorphism, then there is a  $t : B \to J$  such that  tg = f .

Suppose  $0 \to A \to B \to C \to 0$  is an almost split sequence with  A  in  $\text{Art}_I(\Lambda)$  and  C  in  $\text{Noeth}_P(\Lambda)$ .  Then  B  is in  Noeth($\Lambda$)  if and only if  A  is of finite length while  B  is in  Art($\Lambda$)  if and only if  C  is of finite length.  Thus if  $\Lambda$  is not Artinian, then  B  will, in general, not be in either  Noeth($\Lambda$)  or  Art($\Lambda$) .  This fact suggests introducing the full subcategory  Arno($\Lambda$)  of  Mod($\Lambda$)  consisting of all $\Lambda$-modules  M  such that there is an exact sequence  $0 \to M' \to M \to M'' \to 0$  with  M'  in  Art($\Lambda$)  and  M''  in  Noeth($\Lambda$) .  We now point out certain basic properties of the category  Arno($\Lambda$)  which we need in these lectures.

Proposition 3.5.  The full subcategory  Arno($\Lambda$)  of  Mod($\Lambda$)  has the following properties.

(a)  Suppose  $0 \to M' \to M \to M'' \to 0$  is an exact sequence in  Mod($\Lambda$)  Then  M  is in  Arno($\Lambda$)  if and only if  M'  and  M''  are in  Arno($\Lambda$) .

(b)  A module  M  is in  Arno($\Lambda$)  if and only if  D(M)  is in  Arno($\Lambda^{\text{op}}$) .

(c)  If  M  is in  Arno($\Lambda$) , then the natural morphism  $M \to D^2(M)$  is an isomorphism.

(d)  The functor  $D : \text{Mod}(\Lambda) \to \text{Mod}(\Lambda^{\text{op}})$  induces a duality of categories  $D : \text{Arno}(\Lambda) \rightarrow \text{Arno}(\Lambda^{\text{op}})$ .

As a consequence of the last two results we obtain the following proposition.

**Proposition 3.6.**  Let  $0 \rightarrow A \rightarrow B \rightarrow C \rightarrow 0$  be an exact sequence with A  and  C  in  $\text{Art}(\Lambda)$  and  $\text{Noeth}(\Lambda)$ , respectively.  Then the following are equivalent:

(a)  $0 \rightarrow A \rightarrow B \rightarrow C \rightarrow 0$  is an almost split sequence,

(b)  $0 \rightarrow D(C) \rightarrow D(B) \rightarrow D(A) \rightarrow 0$  is an almost split sequence of $\Lambda^{\text{op}}$-modules.

Our final aim in this section is to strengthen the criterion given in Proposition 3.4 for an exact sequence to be an almost split sequence.  However, in order to state this result, it is necessary to recall the notion from commutative ring theory of the **support** of a module.  Suppose  $\Lambda$  is an R-algebra and  M  is  $\Lambda$-module.  Then  M  is also an R-module and we define the **support of**  M  ( supp(M) ) to be the set of prime ideals  $\underline{p}$  of  R  such that  $M_{\underline{p}} \neq (0)$ .  The following properties of the support of modules are well-known (see [7]).

(a)  If  $0 \rightarrow M' \rightarrow M \rightarrow M'' \rightarrow 0$  is an exact sequence of $\Lambda$-modules, then  $\text{supp}(M) = \text{supp}(M') \cup \text{supp}(M'')$ .

(b)  A $\Lambda$-module is different from zero if and only if  $\underline{m}$ , the maximal ideal of  R , is in  supp(M) .

(c)  If  M  is a nonzero finitely generated $\Lambda$-module, then  M  has finite length if and only if  supp(M) = {$\underline{m}$} .

(d)  supp(M) = $\cup$ supp($M_i$)  where  {$M_i$}$_{i \in I}$  is the family of finitely generated submodules of  M .

(e)  If  M  is a nonzero module in  Art($\Lambda$) , then  supp(M) = {$\underline{m}$} .

(f)  $0 \to A \to B \to C \to 0$  is exact with  A  in  Art($\Lambda$)  and  C $\neq$ (0) , the  supp(B) = supp(C) .

Combining these properties of the support with the existence of almost split sequences, we obtain the following proposition.


<u>Proposition 3.7</u>.  (a)  Suppose  C  is an indecomposable module in

Noeth$_p$($\Lambda$)  and  f : X $\to$ C  is a morphism in  Mod($\Lambda$)  which

is not a splittable epimorphism.  Then, there is a factoriza-

tion  X $\xrightarrow{\ g\ }$ Y $\xrightarrow{\ h\ }$ C  of  f  satisfying:

(i)  Y  is in  Arno($\Lambda$)  and  supp(Y) $\subset$ supp(C) ,

(ii)  g  is an epimorphism,

(iii)  h  is not a splittable epimorphism.

(b)  Suppose  A  is an indecomposable module in  Art$_I$($\Lambda$)  and

f : A $\to$ Y  is a morphism in  Mod($\Lambda$)  which is not a splittable

monomorphism.  Then, there is a factorization  A $\xrightarrow{\ g\ }$ X $\xrightarrow{\ h\ }$ Y

of  f  satisfying:

(i)  X  is  Arno($\Lambda$)  and  supp(X) $\subset$ supp(TrD(A)) ,

(ii)  h  is a monomorphism,

(iii)  g  is not a splittable monomorphism.


Applying this result concerning the support of modules with

Proposition 3.4, we obtain our final result of this section.


**Theorem 3.8.** For a nonsplittable exact sequence $0 \longrightarrow A \xrightarrow{g} B \xrightarrow{h}$
$C \longrightarrow 0$ with $A$ and $C$ indecomposable $\Lambda$-modules in $\mathrm{Art}_I(\Lambda)$ and
$\mathrm{Noeth}_P(\Lambda)$ , respectively, the following are equivalent:

  (a)  $0 \longrightarrow A \xrightarrow{g} B \xrightarrow{h} C \longrightarrow 0$ is an almost split sequence.

  (b)  If $X$ is in $\mathrm{Noeth}(\Lambda)$ with $\mathrm{supp}(X) \subset \mathrm{supp}(C)$ and $f : X \to$
        $C$ is not a splittable epimorphism, then there is a morphism
        $s : X \to B$ such that $hs = f$ .

  (c)  If $Y$ is in $\mathrm{Art}(\Lambda)$ and $f : A \to Y$ is not a splittable
        monomorphism, then there is a $t : B \to Y$ such that $f = tg$ .


**§4. Some Applications.** In this section we give some applications of our
previous results to the study of modules over R-algebras $\Lambda$ where $R$
is a complete local ring. To this end it is convenient to make the
following definition.

We say that a $\Lambda$-module $A$ is **proArtin** if it is isomorphic to the
inverse limit of Artin modules over $\Lambda$ . It is not difficult to see
that a $\Lambda$-module $A$ is proArtin if and only if $A \simeq D(X)$ for some
module $X$ over $\Lambda^{op}$ .

Many of our applications of the previous material depend on the
following useful result.


**Lemma 4.1.** Suppose $0 \longrightarrow A \xrightarrow{g} B \xrightarrow{h} C \longrightarrow 0$ is a nonsplittable exact sequence in $\text{Mod}(\Lambda)$ . If  $A$  is a proArtin module, then there is

an indecomposable  $C'$  of  $C$  in  $\text{Noeth}(\Lambda)$  such that the exact sequence

$$0 \to A \to h^{-1}(C') \to C' \to 0$$

is __not__ splittable.


As a consequence of Lemmas 3.3 and 4.1 we obtain the following

proposition.


__Proposition 4.2.__  Suppose  $A$  is a proArtin  $\Lambda$-module.  An exact sequence

$0 \longrightarrow A \xrightarrow{g} B \xrightarrow{h} C \longrightarrow 0$  in  $\text{mod}(\Lambda)$  is splittable if given any

epimorphism  $f : A \to A'' \to 0$  with  $A''$  an indecomposable module in

$\text{Art}(\Lambda)$ , there is a morphism  $s : B \to A''$  such that  $f = sg$ .


As an interesting special case of this proposition we have the next

corollary.


__Corollary 4.3.__  Suppose  $A$  is in  $\text{Noeth}(\Lambda)$ .  Then,

    (a)  An exact sequence  $0 \longrightarrow A \xrightarrow{g} B \xrightarrow{h} C \longrightarrow 0$  in  $\text{Mod}(\Lambda)$

        is splittable if for each integer  $i > 0$ , there is a  $\Lambda$-morphism

        $s_i : B \to A/\underline{m}^i A$  such that  $s_i g : A \to A/\underline{m}^i A$  is the canonical

        epimorhpism.

    (b)  Let  $A \to \prod_{i>0} A/\underline{m}^i A$  be the morphism induced by the canonical

        epimorphisms  $A \to A/\underline{m}^i A$ .  Then  $A \to \prod A/\underline{m}^i A$  is a splittable

        monomorphism.

As another consequence of Proposition 4.2 we have the following
corollary.

Corollary 4.4.   (a)   If  A  is a proArtin $\Lambda$-module such that there is

        an almost split sequence  $0 \to A \to B \to C \to 0$ , then  A  is in

        Art$(\Lambda)$ .

    (b)   Suppose  A  is in  Noeth$(\Lambda)$ .  Then there is an almost split

        sequence  $0 \to A \to B \to C \to 0$  if and only if  A  is an indecom-

        posable noninjective $\Lambda$-module of finite length.

This last result shows that there are noninjective $\Lambda$-modules  A
with  $\text{End}_\Lambda(A)$  local such that there is no almost split sequence  $0 \to A \to$
$B \to C \to 0$  in  Mod$(\Lambda)$ .  It would be interesting to know precisely for
which $\Lambda$-modules  A  with  $\text{End}_\Lambda(A)$  local there are almost split sequences
$0 \to A \to B \to C \to 0$ .  In particular, it would be interesting to have
examples of  nonArtin modules  A  such that there are almost split
sequences  $0 \to A \to B \to C \to 0$ .

    It is not difficult to show that the following result holds for
arbitrary rings (not necessarily R-algebras).

Proposition 4.5.  Suppose  $\Gamma$  is an arbitrary ring and  A  is a finitely
generated $\Gamma$-module with  $\text{End}_\Gamma(A)$  a local ring.  If  $f_i : X_i \to A$  is a
family of morphisms such that the induced morphism  $\amalg\, X_i \to A$  is a
splittable epimorphism, then some  $f_i : X_i \to A$  is a splittable epimorphism.

It is therefore natural to ask if there is an analogous result for products. A partial answer for R-algebras with $R$ a complete local ring is given in the following proposition.

**Proposition 4.6.** Let $\Lambda$ be an R-algebra with $R$ a complete local ring. Suppose $A$ is an indecomposable $\Lambda$-module in $\mathrm{Art}(\Lambda)$. If the family of morphisms $u_i : A \to X_i$ in $\mathrm{Mod}(\Lambda)$ has the property that the induced morphism $A \to \Pi X_i$ is a splittable monomorphism, then some $u_i : A \to X_i$ is a splittable monomorphism.

This result was proven in [1] in the case where $R$ is Artin. The same proof works when $R$ is a complete local ring since it depends only on the existence of an almost split sequence $0 \to A \to B \to C \to 0$ and not on the nature of the rings involved.

Applying Corollary 4.3 together with Proposition 4.6 we obtain the following characterization of modules of finite length.

**Corollary 4.7.** Let $A$ be an indecomposable module in $\mathrm{Noeth}(\Lambda)$ ( $\Lambda$ an R-algebra with $R$ a complete local ring). Then $A$ has finite length if and only if given any family $u_i : A \to X_i$ of monomorphisms in $\mathrm{Noeth}(\Lambda)$ such that the induced morphism $A \to \Pi X_i$ is a splittable monomorphism, then some $u_i : A \to X_i$ is a splittable monomorphism.

**§5.   Lattices and Orders.** Suppose that $\Lambda$ is an R-algebra with $R$ a complete local ring.  We saw in the previous sections that if  C  is

an indecomposable $\Lambda$-module in  $\text{Noeth}_p(\Lambda)$ , then there is an almost split

sequence  $0 \to A \to B \to C \to 0$  with  A  an indecomposable module in

$\text{Art}_I(\Lambda)$ .  We also saw that if  X  is an indecomposable noninjective

$\Lambda$-module in  $\text{Noeth}(\Lambda)$ , then there does not exist an almost split sequence

$0 \to X \to Y \to Z \to 0$  unless  X  is also in  $\text{Art}(\Lambda)$  and, consequently, of

finite length.  This apparent lack of symmetry can in some instances be

overcome by introducing the notion of an almost split sequence relative

to a subcategory  $\underline{A}$  of  $\text{Mod}(\Lambda)$ .  We will consider this relative notion

only in the case  $\underline{A}$  satisfies certain conditions described below.  The

reason for this is that under these hypotheses many of the uniqueness and

formal properties of almost split sequences given in the previous sec-

tions hold for these relative almost split sequences.

   Suppose  $\Gamma$  is an arbitrary ring and  $\underline{A}$  is a full subcategory of

$\text{Mod}(\Gamma)$  satisfying the following conditions:

   (a)  If  A  is in  $\underline{A}$  and  C  in  $\text{Mod}(\Gamma)$  is isomorphic to  A  then

C  is in  $\underline{A}$ .

   (b)  If  $0 \to C_1 \to C_2 \to C_3 \to 0$  is an exact sequence with  $C_1$  and

$C_3$  in  $\underline{A}$ , then  $C_2$  is in  $\underline{A}$ .

   (c)  If  $C_1$  and  $C_2$  are in  $\text{Mod}(\Gamma)$ , then  $C_1 \amalg C_2$  is in  $\underline{A}$  if

and only if  $C_1$  and  $C_2$  are in  $\underline{A}$ .

   We say that an exact sequence  $0 \longrightarrow C_1 \xrightarrow{g} C_2 \xrightarrow{h} C_3 \longrightarrow 0$  is

an  $\underline{A}$  **almost split sequence** if it satisfies the following:

   (a)  The  $C_i$  are in  $\underline{A}$  for  i = 1,2,3 .

   (b)  The sequence is not splittable.

(c) If $f : X \to C_3$ is a morphism in $\underline{A}$ which is not a splittable epimorphism, then there is an $s : X \to C_2$ such that $hs = f$ .

(d) If $f : C_1 \to Y$ is a morphism in $\underline{A}$ which is not a splittable monomorphism, there is a $t : C_2 \to Y$ such that $tg = f$ .

The reader should observe that $\underline{A} = \text{Mod}(\Gamma)$ satisfies the conditions given above and in this case an exact sequence $0 \to C_1 \to C_2 \to C_3 \to 0$ is an almost split sequence if and only if it is an $\underline{A}$ almost split sequence.

As in the case of ordinary almost split sequences we have the following proposition.


**Proposition 5.1.** Let $\underline{A}$ be a full subcategory of $\text{Mod}(\Gamma)$ satisfying the above conditions. Suppose $0 \to C_1 \to C_2 \to C_3 \to 0$ and $0 \to C_1' \to C_2' \to C_3' \to 0$ are two $\underline{A}$ almost split sequences. Then,

(a) $\text{End}(C_i)$ and $\text{End}(C_i')$ are local rings for $i = 1,3$ .

(b) The following statements are equivalent:

(i) the two sequences are isomorphic,

(ii) $C_1 \approx C_1'$ ,

(iii) $C_3 \approx C_3'$ .


Returning to the situation of an R-algebra $\Lambda$ with $R$ a complete local ring, it is reasonable to ask for which full subcategories $\underline{A}$ of $\text{Mod}(\Gamma)$ satisfying the above conditions do $\underline{A}$ almost split sequences exist? While we have no complete answer to this question, our theory to date shows, among other things, that $\underline{A}$ almost split sequences exist in the following cases.

(a)  R  is Artin and  $\underline{A}$ = Noeth($\Lambda$) .

(b)  R  is a one-dimensional domain with field of quotients  K  and

$\Lambda$  is an R-algebra such that  K $\otimes_R$ $\Lambda$  is semisimple.  $\underline{A}$  is the full sub-

category of  Noeth($\Lambda$)  whose objects are the modules which are torsion-

free when viewed as R-modules.  If  R  is a discrete valuation ring,

then the  $\underline{A}$  just described is the category of lattices over  $\Lambda$ .

(c)  R  is an integrally closed two-dimensional local ring and  $\underline{A}$

is the full subcategory of  Noeth($\Lambda$)  consisting of the **reflexive** R-

modules, i.e., a  C  in  Noeth($\Lambda$)  is in  $\underline{A}$  if and only if the natural

morphism  $C \rightarrow Hom_R(Hom_R(C,R),R)$  is an isomorphism.

In order to explain the general theorem of which the above are

special cases, we need to recall some facts from commutative ring theory

(see [8]).

Let  T  be an arbitrary Noetherian local ring not necessarily

complete with maximal ideal  $\underline{m}$ .  Suppose  M  is a nonzero module in

Noeth(T) .  An element  x  in  $\underline{m}$  is said to be **M-regular** if  x  is

not a zero divisor in  M , i.e.,  xm = 0  implies  m = 0  for any  m

in  M .  A sequence  $x_1,\ldots,x_t$  of elements in  $\underline{m}$  is said to be an

**M-regular** **sequence** if  $x_i$  is  $M/(x_i,\ldots,x_{i-1})$  M-regular for all

i = 1,...,t  where  $(x_1,\ldots,x_{i-1})$  is the ideal in  T  generated by

$x_1,\ldots,x_{i-1}$ .  It is well-known that every M-regular sequence can be

extended to a maximal M-regular sequence and that all maximal M-regular

sequences have the same length, i.e., the same number of elements.  The

length of the maximal M-regular sequences is called the **depth** of  M and is denoted by depth(M) . In general, depth(M) $\leq$ dim(T) where

dim(T) is the Krull dimension of T . We say that a T-module is a

**Cohen-Macauley module** if depth(M) = dim(T) . We denote by $\underline{CM}$(T) the

full subcategory of Mod(T) consisting of the Cohen-Macauley T-modules.

It is well-known that $\underline{A}$ = $\underline{CM}$(T) has the properties (a), (b), (c)

given above plus the following.

(d) If $0 \to C_1 \to C_2 \to C_3 \to 0$ is exact with $C_2, C_3$ in $\underline{CM}$(T) ,

then $C_1$ is also in $\underline{CM}$(T) .

Finally, T is said to be a **Cohen-Macauley ring** if depth(T) = dim(T) .

We recall that if T is a Cohen-Macauley ring and $x_1, \ldots, x_t$ is

a T-regular sequence, then $T/(x, \ldots, x_t)$ is also a Cohen-Macauley

ring. Since all regular local rings are Cohen-Macauley rings, if $x_1, \ldots, x_t$

is a T-regular sequence and T is a regular local ring, then

$T/(x_1, \ldots, x_t)$ is a Cohen-Macauley ring. But it is a rather special sort

of Cohen-Macauley ring known as a **Gorenstein ring**, whose definition we

now recall.

A local ring T is said to be **Gorenstein** if inj(dim$_T$(T)) = n < $\infty$ ,

i.e., if $0 \to T \to I_0(T) \to \cdots \to I_i(T) \to \cdots$ is a minimal injective

resolution of T , then $I_i(T) = 0$ for all i > n and $I_n(T) \neq 0$ .

For a Gorenstein ring T we have that dim(T) = depth(T) = inj(dim(T)) ,

so that all Gorenstein rings are Cohen-Macauley. Moreover,

(a) all regular rings are Gorenstein;

(b) if T is Gorenstein and $x_1, \ldots, x_t$ is a T-regular sequence,

then $T/(x_1, \ldots, x_t)$ is Gorenstein;

(c) if T is Gorenstein, then $T_p$ is Gorenstein for all prime ideals $\underline{p}$ of $T$ .

We shall be mainly interested in Cohen-Macauley modules over

Gorenstein rings.  We now list some of the properties of these

modules we will need.


<u>Proposition 5.2.</u>  Let  $T$  be a Gorenstein ring of dimension  $d$ .

(a)  If  $0 \to M_d \to M_{d-1} \to \cdots \to M_0$  is an exact sequence in

   Noeth(T)  with the  $M_i$  in  $\underline{\underline{CM}}(T)$  for  $i = 0,\ldots,d-1$ ,

   then  $M_d$  is in  $\underline{\underline{CM}}(T)$ .

(b)  If  $M$  is in  $\underline{\underline{CM}}(T)$ , then  $\mathrm{Hom}_T(M,T)$  is in  $\underline{\underline{CM}}(T)$ .

(c)  Each  $M$  in  $\underline{\underline{CM}}(T)$  is reflexive, i.e., the natural morphism

   $M \to \mathrm{Hom}_T(\mathrm{Hom}_T(M,T),T)$  is an isomorphism.

(d)  If  $M$  is in  $\underline{\underline{CM}}(T)$ , then  $\mathrm{Ext}^i_T(M,T) = 0$  for all  $i > 0$ .

(e)  The contravariant functor  $\mathrm{Hom}_T(\ ,T) : \underline{\underline{CM}}(T) \to \underline{\underline{CM}}(T)$  is a

   duality.

(f)  Suppose  $0 \to M_1 \to M_2 \to M_3 \to 0$  is an exact sequence in

   Noeth(T) .  Then,

     (i)  if  $M_1$  and  $M_3$  are in  $\underline{\underline{CM}}(T)$ , then  $M_2$  is in  $\underline{\underline{CM}}(T)$ ,

     (ii)  if  $M_2$  and  $M_3$  are in  $\underline{\underline{CM}}(T)$ , then  $M_1$  is in  $\underline{\underline{CM}}(T)$ ,

     (iii)  if the  $M_i$  are in  $\underline{\underline{CM}}(T)$ , then  $0 \to \mathrm{Hom}_T(M_3,T) \to$

        $\mathrm{Hom}_T(M_2,T) \to \mathrm{Hom}_T(M_1,T) \to 0$  is exact.

(g)  If  $M$  is in  $\underline{\underline{CM}}(T)$ , then  $M_{\underline{p}}$  is in  $\underline{\underline{CM}}(T_{\underline{p}})$  for all prime

   ideals  $\underline{p}$  of  $T$ .


Suppose now that  $R$  is a complete local ring.  Then, there is a complete regular local U such that $U/\underline{a} \approx R$ for some ideal $\underline{a}$ of U . Let $x_1, \ldots, x_t$ be a maximal U-regular sequence contained in $\underline{a}$ . Then $T = U/(x_1, \ldots, x_t)$ is a Gorenstein ring with $\dim(T) = \dim(R)$ . Thus there is a surjective ring morphism $T \to R$ with $T$ a complete Gorenstein with $\dim(T) = \dim(R)$ . While there is no unique way of representing $R$ as a factor of a complete Gorenstein ring $T$ with $\dim(T) = \dim(R)$ , there are properties of this type of representation which are independent of the particular representation chosen. We list some of these in the following proposition.

**Proposition 5.3.** Let $T_1 \to R$ and $T_2 \to R$ be surjective ring morphisms with the $T_i$ complete Gorenstein local rings of the same dimension as $R$ .

    (a) The following are equivalent for $M$ in $Noeth(R)$ :

        (i) $M$ is in $\underline{\underline{CM}}(R)$ ,

        (ii) $M$ is in $\underline{\underline{CM}}(T_1)$ ,

        (iii) $M$ is in $\underline{\underline{CM}}(T_2)$ .

    (b) The functors $\mathrm{Hom}_{T_i}(\ , T_i) = \underline{\underline{CM}}(R) \to \underline{\underline{CM}}(R)$ are isomorphic dualities.

We assume from now on that we have chosen a fixed surjective ring morphism of complete local rings $T \to R$ of the same dimension with $T$ a Gorenstein ring.

We are now in a position to introduce the notion of a lattice over

an R-algebra $\Lambda$ which generalizes the usual notion of lattices of orders
over complete discrete valuation rings.

Let $\Lambda$ be an R-algebra. We say that a $\Lambda$-module $M$ in $\text{Noeth}(\Lambda)$
is <u>Cohen-Macauley</u> if it is a Cohen-Macauley when viewed as an R-module.
We denote the full subcategory of Cohen-Macauley $\Lambda$-modules by $\underline{\underline{CM}}(\Lambda)$ .
A $\Lambda$-module $M$ is said to be a **lattice** if $M$ is in $\underline{\underline{CM}}(\Lambda)$ and $M_{\underline{p}}$ and
$\text{Hom}_T(M,T)_{\underline{p}}$ are projective $\Lambda_{\underline{p}}$ and $\Lambda_{\underline{p}}^{\text{op}}$-modules for all nonmaximal prime
ideals $\underline{p}$ of $R$ . We denote by $\underline{\underline{latt}}(\Lambda)$ the full subcategory of
$\text{Noeth}(\Lambda)$ whose objects are $\Lambda$-lattices. Moreover, we say that $\Lambda$ is
an **R-order** if $\Lambda$ is in $\underline{\underline{latt}}(\Lambda)$ .

We now summarize some properties of $\underline{\underline{latt}}(\Lambda)$ we will need.


<u>Proposition 5.4.</u> Let $\Lambda$ be an R-order with $\dim(R) = d$ . Then,

    (a) For $M$ in $\text{Noeth}(\Lambda)$ the following are equivalent:

        (i) $M$ is in $\underline{\underline{latt}}(\Lambda)$ ,

        (ii) $M$ is in $\underline{\underline{CM}}(\Lambda)$ and $M_{\underline{p}}$ is $\Lambda_{\underline{p}}$-projective for all
           nonmaximal prime ideals $\underline{p}$ of $R$ ,

        (iii) $M$ is in $\underline{\underline{CM}}(\Lambda)$ and $\text{Hom}_T(M,T)$ is in $\underline{\underline{latt}}(\Lambda^{\text{op}})$ .

    (b) The functor $\text{Hom}_T(\ ,T) : \underline{\underline{latt}}(\Lambda) \to \underline{\underline{latt}}(\Lambda^{\text{op}})$ is a duality.

    (c) Suppose $0 \to M_1 \to M_2 \to M_3 \to 0$ is an exact sequence in
    $\text{Noeth}(\Lambda)$ .

        (i) If the sequence splits, then $M_2$ is in $\underline{\underline{latt}}(\Lambda)$ if
           and only if $M_1$ and $M_3$ are in $\underline{\underline{latt}}(\Lambda)$ .

        (ii) If $M_1$ and $M_3$ are in $\underline{\underline{latt}}(\Lambda)$ , then $M_2$ is in
           $\underline{\underline{latt}}(\Lambda)$ .

(iii)  If $M_2$ and $M_3$ are in $\underline{\text{latt}}(\Lambda)$ , then $M_1$ is in $\underline{\text{latt}}(\Lambda)$ .

(iv)  If the $M_i$ are in $\underline{\text{latt}}(\Lambda)$ , then $0 \to \text{Hom}_T(M_3,T) \to \text{Hom}_T(M_2,T) \to \text{Hom}_T(M_1,T) \to 0$ is exact with the $\text{Hom}_T(M_i,T)$ in $\underline{\text{latt}}(\Lambda^{\text{op}})$ for all $i$ .

(d)  If $\cdots \longrightarrow P_{i+1} \xrightarrow{\ d_{i+1}\ } P_i \xrightarrow{\ d_i\ } \cdots \longrightarrow P_1 \xrightarrow{\ d_1\ }$

$P_0 \xrightarrow{\ d_0\ } M \longrightarrow 0$ is a projective resolution of $M$ over $\Lambda$ , then $\text{Im}(d_i)$ is a lattice for all $i \geq \dim(R)$ .

Thus we see if we let $\underline{A}$ be the full subcategory $\underline{\text{latt}}(\Lambda)$ , then we see that $\underline{A}$ satisfies the conditions given in the beginning of this section.

Thus it makes sense to ask if $\underline{\text{latt}}(\Lambda)$ almost split sequences exist. This question is answered in the following theorem.

Theorem 5.3.  Suppose $\Lambda$ is an R-order. Then,

(a)  if $A$ is an indecomposable $\Lambda$-lattice such that $\text{Hom}_T(A,T)$ is not $\Lambda^{\text{op}}$-projective, then there is a $\underline{\text{latt}}(\Lambda)$ almost split sequence $0 \to A \to B \to C \to 0$ ;

(b)  if $C$ is a nonprojective indecomposable $\Lambda$-lattice, then there is a $\underline{\text{latt}}(\Lambda)$ almost split sequence $0 \to A \to B \to C \to 0$ .

Unlike previous results, the proof of this theorem is not analogous to already published results concerning Artin algebras. For this reason we outline the proof of this theorem which will also give a description

of how the ends of a $\underline{latt}(\Lambda)$ almost split sequence are related.

To begin with we observe that it follows from the duality

$Hom_T(\ ,T) : \underline{latt}(\Lambda) \to \underline{latt}(\Lambda^{op})$ that parts (a) and (b) of the theorem

are equivalent, so it suffices to establish only one of them. We will

now show that (a) is valid. This proof proceeds in several steps.


<u>Lemma 5.4.</u>  Let $\Lambda$ be an R-algebra ( $\Lambda$ not necessarily an R-order) and

assume that $\dim(R) = d \geq 0$ . Further, let $0 \to T \to I_0 \to I_1 \to \cdots \to I_d \to 0$

be a minimal injective resolution of $T$ over $T$ .

   (a)  If $M$ is in $\underline{latt}(\Lambda)$ , then $Hom_T(M,I_j)$ is an injective

        $\Lambda^{op}$-module for all $j = 0,\ldots,d - 1$ .

   (b)  For each $M$ in $\underline{latt}(\Lambda)$ , there is an isomorphism of functors

        $Ext^1_{\Lambda^{op}}(\ ,Hom_T(M,I_d)) \cong Ext^{d+1}_{\Lambda^{op}}(\ ,Hom_T(M,T))$ from $Mod(\Lambda)$ to

        abelian groups which is functorial in $M$ , i.e., if $f : M \to N$

        is a morphism in $\underline{latt}(\Lambda)$ , then the induced morphisms

        $Hom_T(N,I_d) \to Hom_T(M,I_d)$ and $Hom_T(N,T) \to Hom_T(M,T)$ give rise

        to the following commutative diagram.

$$
\begin{array}{ccc}
Ext^1_{\Lambda^{op}}(\ ,Hom_T(N,I_d)) & \cong & Ext^{d+1}_{\Lambda^{op}}(\ ,Hom_T(N,T)) \\
\downarrow & & \downarrow \\
Ext^1_{\Lambda^{op}}(\ ,Hom_T(M,I_d)) & \cong & Ext^{d+1}_{\Lambda^{op}}(\ ,Hom_T(M,T))
\end{array}
$$

Proof: (a) Suppose $M$ is in latt$(\Lambda)$ and $0 \leq j \leq d - 1$. If

$d = 0$, there is nothing to prove, so we may as well assume that $d > 0$

Since $I_j$ is an injective T-module, we know that for all $\Lambda^{op}$-modules

$A$ we have an isomorphism $\text{Ext}^1_{\Lambda^{op}}(A,\text{Hom}_T(M,I_j)) \cong \text{Hom}_T(\text{Tor}^\Lambda_1(A,M),I_j)$ .

Since $M$ is in latt$(\Lambda)$ , we know that $M_{\underline{p}}$ is $\Lambda_{\underline{p}}$-projective for all

nonmaximal primes $\underline{p}$ in $R$ . Thus for each $A$ in Noeth$(\Lambda^{op})$ , we

know that $\text{Tor}^\Lambda_1(A,M)$ has finite length over $R$ and hence also over $T$

But the fact that $T$ is Gorenstein implies that $\text{Hom}_T(X,I_j) = 0$ for

all T-modules of finite length $X$ for all $j = 0,\ldots,d - 1$ . Thus

$\text{Ext}^1_{\Lambda^{op}}(A,\text{Hom}_T(M,I_j)) = 0$ for $j = 0,\ldots,d - 1$ for all $A$ in Noeth$(\Lambda^{op}$

which shows that $\text{Hom}_T(M,I_j)$ is $\Lambda^{op}$-injective for $j = 0,\ldots,d - 1$ .

(b) Suppose $M$ and $N$ are in latt$(\Lambda)$ and $f : M \rightarrow N$ is a

morphism. We then have the exact commutative diagram of $\Lambda^{op}$-modules

$$
\begin{array}{ccccccccc}
0 & \rightarrow & \text{Hom}_T(N,T) & \rightarrow & \text{Hom}_T(N,I_0) & \rightarrow & \cdots & \rightarrow & \text{Hom}_T(N,I_d) & \rightarrow & 0 \\
& & \downarrow{\scriptstyle\text{Hom}_T(f,T)} & & \downarrow{\scriptstyle\text{Hom}_T(f,I_0)} & & & & \downarrow{\scriptstyle\text{Hom}_T(f,I_d)} & & \\
0 & \rightarrow & \text{Hom}_T(M,T) & \rightarrow & \text{Hom}_T(M,I_0) & \rightarrow & \cdots & \rightarrow & \text{Hom}_T(M,I_d) & \rightarrow & 0
\end{array}
$$

That the diagram commutes is obvious. That the rows are exact follows

from the fact that $\text{Ext}^i_T(X,T) = 0$ for $i > 0$ for all Cohen-Macauley

modules $X$ since $T$ is Gorenstein. In particular, $\text{Ext}^i_T(M,T) = 0 =$

$\text{Ext}^i_T(N,T)$ for all $i > 0$ for $M$ and $N$ in latt$(\Lambda)$ .

Now by (a), we know that $\text{Hom}_T(N,I_j)$ and $\text{Hom}_T(M,I_j)$ are injective

$\Lambda^{op}$-modules for $j = 0,\ldots,d - 1$ . Therefore by standard dimension

shift arguments we obtain the following commutative exact diagram

$$0 \rightarrow \mathrm{Ext}^1_{\Lambda^{op}}(\ ,\mathrm{Hom}_T(N,I_d)) \rightarrow \mathrm{Ext}^{d+1}_{\Lambda^{op}}(\ ,\mathrm{Hom}_T(N,T)) \rightarrow 0$$

$$0 \rightarrow \mathrm{Ext}^1_{\Lambda^{op}}(\ ,\mathrm{Hom}_T(M,I_d)) \rightarrow \mathrm{Ext}^{d+1}_{\Lambda^{op}}(\ ,\mathrm{Hom}_T(M,T)) \rightarrow 0$$

which is our desired result.

We now apply this lemma to obtain the following proposition.

**Proposition 5.5.** Let $\Lambda$ be an R-order and $d = \dim(R)$. Suppose $A$ is an indecomposable $\Lambda$-lattice such that $\mathrm{Hom}_T(A,T)$ is not $\Lambda^{op}$-projective. Let $V = \mathrm{Tr}(\mathrm{Hom}_T(A,T))$ and $0 \rightarrow K \rightarrow P_{d-1} \rightarrow \cdots \rightarrow P_0 \rightarrow P_{-1} \rightarrow 0$ an exact sequence of $\Lambda$-modules where $P_{-1} = V$ and $P_{d-1} \rightarrow \cdots \rightarrow P_0 \rightarrow V \rightarrow 0$ the beginning of a minimal projective resoltuion of $V$. Then,

    (a) $V$ is a nonprojective inedcomposable module in $\mathrm{Noeth}(\Lambda)$ such that $V_{\underline{p}}$ is $\Lambda_{\underline{p}}$-projective for all nonmaximal prime ideals $\underline{p}$ of $R$.

    (b) $K$ is a $\Lambda$-lattice having the following properties:

        (i) for each $X$ in $\underline{\mathrm{latt}}(\Lambda)$, there is an isomorphism $\mathrm{Ext}^1_\Lambda(K,X) \cong \mathrm{Ext}^1(V,\mathrm{Hom}_T(\mathrm{Hom}_T(X,T),I(T/\mathrm{rad}(T)))$ which is functorial in $X$;

        (ii) there is a nonsplittable exact sequence $0 \longrightarrow A \xrightarrow{g} B \xrightarrow{h} K \longrightarrow 0$ of $\Lambda$-lattices with the property that if $f : A \rightarrow M$ is a morphism of $\Lambda$-lattices which is not a splittable monomorphism, then there is a morphism $s : B \rightarrow M$ such that $sg = f$.

Proof: (a) Since $\text{Hom}_T(A,T)$ is an indecomposable nonprojective $\Lambda^{op}$-module, it follows that $V = \text{Tr}(\text{Hom}_T(A,T))$ is an indecomposable non-projective $\Lambda$-module. The fact that $A$ is a $\Lambda$-lattice implies that $\text{Hom}_T(A,T)_{\underline{p}}$ is $(\Lambda^{op})_{\underline{p}}$-projective for all nonmaximal prime ideals $\underline{p}$ of $R$. This, in turn, implies that $V_{\underline{p}}$ is $\Lambda_{\underline{p}}$-projective for all non-maximal prime ideals $\underline{p}$ of $R$.

(b) Since $\Lambda$ is an R-order, we know that all projective $\Lambda$-modules are Cohen-Macauley. From this it follows that $K$ is Cohen-Macauley. The fact that $V_{\underline{p}}$ is $\Lambda_{\underline{p}}$-projective for all nonmaximal prime ideals $\underline{p}$ of $R$, implies that $K_{\underline{p}}$ is also $\Lambda_{\underline{p}}$-projective for all non-maximal prime ideals $\underline{p}$ of $R$. Since $\Lambda$ is an R-order, this implies that $K$ is a $\Lambda$-lattice.

(i) We first recall that for all $X$ in $\underline{\text{latt}}(\Lambda)$, the canonical morphism $X \rightarrow \text{Hom}_T(\text{Hom}_T(X,T),T)$ is an isomorphism. This part of the proposition now follows from part (b) of Lemma 5.4 and the observation that $I_d = I(T/\text{rad}(T))$ since $T$ is Gorenstein.

(ii) Let $\phi_X : \text{Ext}^1_\Lambda(V,\text{Hom}_T(\text{Hom}_T X,T),I(T/\text{rad}(T))) \rightarrow \text{Ext}^1(K,X)$ be the functorial isomorphism in (i). Let $x$ in $\text{Ext}^1(K,A)$ be $\phi_A(y)$ where $y$ is an almost split sequence in $\text{Ext}^1_\Lambda(V,\text{Hom}_T(\text{Hom}_T(A,T) , I(T/\text{rad}(T)))$ (remember that $\text{Tr}(V) = \text{Hom}_T(A,T)$). Using various dualities, it is not difficult to see that the extension $0 \rightarrow A \rightarrow B \rightarrow K \rightarrow 0$ corresponding to the element $x$ in $\text{Ext}^1_\Lambda(K,A)$ has the desired properties.

Next we observe that the duality $T : \underline{\text{latt}}(\Lambda) \rightarrow \underline{\text{latt}}(\Lambda^{op})$ induces an isomorphism $\text{End}_\Lambda(A) \rightarrow \text{End}_{\Lambda^{op}}(\text{Hom}_T(A,T))^{op}$. Thus, if $\Lambda$ is an R-order then $\text{End}(A)/\underline{a} \cong \underline{\text{End}}(\text{Hom}_T(A,T))^{op}$ where $\underline{a}$ is the ideal of

$\text{End}(A)$ consisting of all endomorphism $f : A \to A$ which factor through

$\Lambda$-lattices $X$ such that $\text{Hom}_T(X,T)$ is $\Lambda^{op}$-projective. Now

$\text{Ext}_\Lambda^1(V,\text{Hom}_T(\text{Hom}_T(A,T),I(T/\text{rad}(T)))$ is isomorphic to the injective enve-

lope of the unique simple $\underline{\text{End}}(\text{Hom}_T(A,T))$ module over $\underline{\text{End}}(\text{Hom}_T(A,T))$ .

Thus $\text{Ext}_\Lambda^1(K,A)$ is isomorphic to the injective envelope over $(\text{End}(A))/\underline{a}$

of the unique simple $(\text{End}(A))/\underline{a}$-module. In particular, $\text{Ext}_\Lambda^1(K,A)$ is

an indecomposable $(\text{End}(A))/\underline{a}$-module. Thus if $K = \amalg K_i$ with the $K_i$

indecomposable lattices, there is only one $i$ such that $\text{Ext}_\Lambda^1(K_i,A) \neq 0$ .

Thus the inclusion $K_i \to K$ induces an isomorphsim $\text{Ext}_\Lambda^1(K,A) \to \text{Ext}_\Lambda^1(K_i,A)$ .

It then follows from Proposition 5.5, that there is a nonsplittable

exact sequence $0 \longrightarrow A \overset{g}{\longrightarrow} B \overset{h}{\longrightarrow} C \longrightarrow 0$ with $C = K_i$ such that

if $f : A \to X$ is a morphism in $\underline{\text{latt}}(\Lambda)$ which is not a splittable mono-

morphism, then there is a $s : B \to X$ such that $sg = f$ . Since $C$ is an

indecomposable lattice, this implies that $0 \longrightarrow A \overset{g}{\longrightarrow} B \overset{h}{\longrightarrow} C \longrightarrow 0$

is a $\underline{\text{latt}}(\Lambda)$ almost split sequence. We summarize these observations

in the next proposition.

Proposition 5.6. Let $\Lambda$ be an R-order with $\dim(R) = d$ , $A$ an in-

decomposable $\Lambda$-lattice such that $\text{Hom}_T(A,T)$ is not $\Lambda^{op}$-projective. Let

$V = \text{Tr}(\text{Hom}_T(A,T))$ and let $0 \to K \to P_{d-1} \to \cdots \to P_1 \to P_0 \to P_{-1} \to 0$ be

exact where $P_{-1} = V$ and $P_{d-1} \to \cdots \to P_0 \to V \to 0$ is the beginning of

a minimal projective resolution of $V$ . Then,

(a) $K$ is a lattice which has a unique (up to isomorphism)

indecomposable summand $C$ such that the induced morphism

$Ext_\Lambda^1(K,A) \to Ext_\Lambda^1(C,A)$ is an isomorphism.

(b) $Ext_\Lambda^1(C,A)$ is the injective envelope over $(End(A))/\underline{a}$ of the simple $(End(A))/\underline{a}$-module. Where $\underline{a}$ is the ideal consisting of all $f : A \to A$ which factors through lattices $X$ with the property that $Hom_T(X,T)$ is $\Lambda^{op}$-projective.

(c) $(End(A))/\underline{a}$ is an R-module of finite length hence so is $Ext_\Lambda^1(C,A)$ .

(d) If $0 \to A \to B \to C \to 0$ is a nonzero element of the $soc(Ext_\Lambda^1(C,A))$ over $(End(A))/\underline{a}$ , then $0 \to A \to B \to C \to 0$ is a $\underline{latt}(\Lambda)$ almost split sequence.

Proof: Everything except part (c) has been discussed before the statement of the proposition. Part (c) is an immediate consequence of the fact that $A$ is a lattice and so $A_{\underline{p}}$ and $Hom_T(A,T)_{\underline{p}}$ are $\Lambda_{\underline{p}}$-projective for all nonmaximal prime ideals $\underline{p}$ of $R$ .

This proposition obviously finishes the proof of Theorem 5.3.

As the final result of this section we state the following easily verified consequence of Proposition 5.6.

Proposition 5.7. Let $\Lambda$ be an R-order with $d = dim(R)$ . Suppose $0 \to A \to B \to C \to 0$ is a $\underline{latt}(\Lambda)$ almost split sequence. Then,

(a) $0 \to Hom_T(C,T) \to Hom_T(B,T) \to Hom_T(A,T) \to 0$ is a $\underline{latt}(\Lambda)$ almost split sequence;

(b) $Ext_\Lambda^1(C,A) \cong Ext_{\Lambda^{op}}^1(Hom_T(A,T),Hom_T(C,T))$ as $End(C)^{op}$-modules;

(c) $Ext_\Lambda^1(C,A)$ is an injective envelope over $\underline{End(C)}^{op}$ of the unique simple $\underline{\mathrm{End}}(C)^{op}$-module.  Further  $\underline{\mathrm{End}}(C)^{op}$  and

$\mathrm{Ext}^1_\Lambda(C,A)$  are R-modules of finite length;

(d)  let  $0 \to K \to P_{d-1} \to \cdots \to P_0 \to P_{-1} \to 0$  be an exact

sequence of $\Lambda^{op}$-modules with  $\mathrm{Tr}(C) = P_{-1}$  and  $P_{d-1} \to \cdots \to$

$P_0 \to \mathrm{Tr}(C) \to 0$  the beginning of a minimal projective

resolution of  $\mathrm{Tr}(C)$ .  Then there is an indecomposable sum-

mand  L  of  K  such that  A  is isomorphic to  $\mathrm{Hom}_T(L,T)$ .

The following property of lattices over an R-order  $\Lambda$ , which we
state without proof, enables us to give a clearer description of the
relationship between  A  and  C  in a  $\underline{\mathrm{latt}}(\Lambda)$  almost split sequence
$0 \to A \to B \to C \to 0$  than those given in Proposition 5.6 and 5.7.

__Proposition 5.8.__  Let  $\Lambda$  be an R-order with  $\dim(R) = d \geq 0$  and  M  an
indecomposable nonprojective $\Lambda$-lattice.

(a)  $\mathrm{Ext}^i_\Lambda(\mathrm{Tr}(M),\Lambda) = 0$  for  $1 \leq i \leq d$ .

(b)  If  $\cdots \longrightarrow P_d \xrightarrow{f_d} P_{d-1} \longrightarrow \cdots \longrightarrow P_0 \xrightarrow{f_0} \mathrm{Tr}(M) \longrightarrow 0$

is a minimal $\Lambda^{op}$-projective resolution of  $\mathrm{Tr}(M)$ , then  $\mathrm{Im}(f_i)$

for  $i = 0,\ldots,d$  is a nonprojective indecomposable $\Lambda^{op}$-module

with the property  $\underline{\mathrm{End}}(\mathrm{Im}(f_i)) \approx \underline{\mathrm{End}}(\mathrm{Tr}(M))$ .

Combining this result with Propositions 5.6 and 5.7 we obtain the
following theorem.

__Theorem 5.9.__  Let  $\Lambda$  be an R-order with  $\dim(R) = d \geq 0$ .

(a)  Suppose  C  is an indecomposable nonprojective $\Lambda$-lattice and

that  $\cdots \longrightarrow P_d \xrightarrow{f_d} P_{d-1} \longrightarrow \cdots \longrightarrow P_0 \xrightarrow{f_0} \mathrm{Tr}(C) \longrightarrow 0$

is a minimal $\Lambda^{op}$-projective resolution of $\mathrm{Tr}(C)$ . If we let $A = \mathrm{Hom}_T(\mathrm{Im}(f_d),T)$ , then there is a $\underline{\mathrm{latt}}(\Lambda)$ almost split sequence $0 \to A \to B \to C \to 0$ .

(b) Suppose $A$ is an indecomposable $\Lambda$-lattice such that $\mathrm{Hom}_T(A,T)$ is not $\Lambda^{op}$-projective and that $\cdots \longrightarrow Q_d \xrightarrow{g_d} Q_{d-1} \longrightarrow$ $\cdots \longrightarrow Q_0 \xrightarrow{g_0} \mathrm{Tr}(\mathrm{Hom}_T(X,T)) \longrightarrow 0$ is a minimal $\Lambda$-projec resolution of $\mathrm{Tr}(\mathrm{Hom}_T(A,T))$ . If we let $C = \mathrm{Im}(g_d)$ , then th is a $\underline{\mathrm{latt}}(\Lambda)$ almost split sequence $0 \to A \to B \to C \to 0$ .

§6. **Examples**. We begin by pointing out when a complete local ring $R$ is an R-order. The following characterization of such rings $R$ is a consequence of the well-known and easily verified fact that if $T \to U$ is a surjective ring morphism of local rings (not necessarily complete) of the same dimension with T-Gorenstein, then $U$ is Gorenstein if and only if $U \cong \mathrm{Hom}_T(U,T)$ or equivalently, $\mathrm{Hom}_T(U,T)$ is a projective U-module.

Proposition 6.1. Let $R$ be a complete local ring. Then $R$ is an R-order if and only if

(a) $R$ is Cohen-Macauley, and

(b) $R_{\underline{p}}$ is Gorenstein for all nonmaximal prime ideals $\underline{p}$ of $R$ .

If $R$ is an R-order, then $\underline{\mathrm{latt}}(R)$ consists of the Cohen-Macauley R-modules as such that $M_{\underline{p}}$ is $R_{\underline{p}}$-free for all nonmaximal prime ideals $\underline{p}$ of $R$ .

On the basis of this characterization of when  R  is an R-order it
is easy to establish the following examples of complete local rings  R
which are R-orders.

Proposition 6.2.  Let  R  be a complete local domain.  Then  R  is an R-
order in the following cases:

   (a)  dim(R) = 1 .  In this case the R-lattices consist of the torsion-
        free R-modules;

   (b)  R  is an integrally closed domain of dimension  2 .  Then
        latt(R)  consists of the reflexive R-modules;

   (c)  R  is Cohen-Macauley and  $R_{\underline{p}}$  is regular for all nonmaximal
        ideals  $\underline{p}$  of  R .  Then  latt(R)  consists of the Cohen-
        Macauley R-modules.

We now turn our attention to various examples of R-orders  Λ  where  Λ
need not be commutative.

Proposition 6.3.  Let  R  be a complete, integrally closed local domain
with field of quotients  K  and suppose  Σ  is a semisimple  K-algebra.

   (a)  If  dim(R) = 1 , then  R  is a complete discrete valuation ring
        and every classical R-order  Λ  of  R  in  Σ  (i.e.,  Λ  is a
        subring of  Σ  containing  R  such that  KΛ = Σ ) is an R-order
        in the sense of this paper.  If  Λ  is an R-order, then  latt(Λ)
        consists of the classical lattices, i.e.,  M  is a Λ-lattice
        if and only if  M  is a free R-module.  (K. Roggenkamp and J.

Schmidt have given a different proof that $\underline{latt}(\Lambda)$ almost split sequences exist in this case.)

(b)   If   $\dim(R) = 2$ , then every maximal R-order   $\Lambda$   in   $\Sigma$ , is an R-order in the sense of this paper. If   $\Lambda$   is a maximal R-order, then   $\underline{latt}(\Lambda)$   consists of the R-reflexive $\Lambda$-modules, i.e.,   M   is in   $\underline{latt}(\Lambda)$   if and only if the natural morphism $M \rightarrow \operatorname{Hom}_R(\operatorname{Hom}_R(M,R),R)$   is an isomorphism.

The rest of this section will be devoted to discussing the importan case of Gorenstein R-algebras which contain   commutative Gorenstein ring as a special case.

Let   R   be a complete local ring of dimension   $d \geq 0$   and let $T \rightarrow R$   be a surjective ring morphism with   T   a Gorenstein local ring. We say that an R-algebra is a $\underline{\text{Gorenstein R-algebra}}$ if   $\operatorname{depth}(\Lambda) = \dim(R)$ and   $\Lambda \cong \operatorname{Hom}_T(\Lambda,T)$   as a two-sided $\Lambda$-module. It is obvious that a Gorenstein R-algebra is an R-order. We now show that some of our considerations in the last section become somewhat simpler for Gorenstein R-algebras. We begin with some preliminaries concerning Gorenstein algebras.

$\underline{\text{Proposition 6.4.}}$   Suppose   $\Lambda$   is a Gorenstein R-algebra. Then,

(a)   $\Lambda^{op}$   is a Gorenstein R-algebra;

(b)   $\Lambda$   is an R-order;

(c)   the functors   $\operatorname{Hom}_\Lambda(\ ,\Lambda), \operatorname{Hom}_T(\ ,T) : \operatorname{Noeth}(\Lambda) \rightarrow \operatorname{Noeth}(\Lambda^{op})$ are isomorphic;

(d)  the natural morphism  $M \to \mathrm{Hom}_{\Lambda^{op}}(\mathrm{Hom}_\Lambda(M,\Lambda),\Lambda)$  is an isomorphism

for all  M  in  <u>latt</u>($\Lambda$) ;

(e)  if  M  is in  <u>latt</u>($\Lambda$) , then  $\mathrm{Ext}^i_\Lambda(M,\Lambda) = 0 = \mathrm{Ext}^i_{\Lambda^{op}}(\mathrm{Hom}_\Lambda(M,\Lambda),\Lambda^{op})$

for all  i > 0 ;

(f)  suppose  M  is in  <u>latt</u>($\Lambda$)  and has no nonzero projective sum-

mands and that  $0 \to \Omega^1(M) \to P \to M \to 0$  is exact with  $P \to M \to 0$

a projective cover.  Then,

  (i)  $\Omega^1(M)$  has no nonzero projective summands and is in

       <u>latt</u>($\Lambda$) ,

  (ii)  $0 \to \mathrm{Hom}_\Lambda(M,\Lambda) \to \mathrm{Hom}_\Lambda(P,\Lambda) \to \mathrm{Hom}_\Lambda(\Omega^1(M),\Lambda) \to 0$  is an

       exact sequence of  $\Lambda^{op}$-modules with the properties:

       $\mathrm{Hom}_\Lambda(M,\Lambda)$  and  $\mathrm{Hom}_\Lambda(\Omega^1(M),\Lambda)$  are in  <u>latt</u>($\Lambda^{op}$)  and

       have no nonzero projective summand and  $\mathrm{Hom}_\Lambda(P,\Lambda) \to$

       $\mathrm{Hom}_\Lambda(\Omega^1(M),\Lambda) \to 0$  is a  $\Lambda^{op}$-projective cover,

  (iii)  <u>End</u>(M)  and  <u>End</u>($\Omega^1(M)$)  are isomorphic R-algebras,

  (iv)  M  is indecomposable if and only if  $\Omega^1(M)$  is

       indecomposable;

(g)  if  M  is an indecomposable nonprojective $\Lambda$-module in  <u>latt</u>($\Lambda$) ,

and  $0 \to \Omega^2(M) \to P_1 \to P_0 \to M \to 0$  is exact with  $P_1 \to P_0 \to M \to$

0  a minimal projective presentation of  M , then,

  (i)  Tr(M) = $\mathrm{Hom}_\Lambda(\Omega^2(M),\Lambda)$  and is therefore in  <u>latt</u>($\Lambda$) ,

  (ii)  $\mathrm{Hom}_T(\mathrm{Tr}(M),T) \cong \mathrm{Hom}_\Lambda(\mathrm{Tr}(M),\Lambda) = \Omega^2(M)$ .


<u>Proof</u>:  (a) and (b) are trivial.

  (c)  Let  M  be in  Noeth($\Lambda$)  then  $\mathrm{Hom}_\Lambda(M,\Lambda) \cong \mathrm{Hom}_\Lambda(M,\mathrm{Hom}_T(\Lambda,T)) \cong \mathrm{Hom}_T(M,T)$

by the usual associative laws.

(d)  follows from (c) and the fact that since  M  is in  $\underline{latt}(\Lambda)$

the natural morphism  $M \to \text{Hom}_T(\text{Hom}_T(M,T),T)$  is an isomorphism.

(e)  Let  M  be in  $\underline{latt}(\Lambda)$  and let  $0 \to \Omega^1(M) \to P \to M \to 0$  be

exact with  $P \to M \to 0$  a projective cover.  Since  M  and  P  are in

$\underline{latt}(\Lambda)$ ,  $\Omega^1(M)$  is in  $\underline{latt}(\Lambda)$  and the sequence  $0 \to \text{Hom}_T(M,T) \to$

$\text{Hom}_T(P,T) \to \text{Hom}_T(\Omega^1(M),T) \to 0$  is exact.  Hence the sequence  $0 \to$

$\text{Hom}_\Lambda(M,\Lambda) \to \text{Hom}_\Lambda(P,\Lambda) \to \text{Hom}_\Lambda(\text{Hom}_\Lambda(\Omega^1(M),\Lambda) \to 0$  is exact which shows

that  $\text{Ext}^1(M,\Lambda) = 0$ .  Since  $\Omega^1(M)$  is in  $\underline{latt}(\Lambda)$ , the same argument

shows that  $0 = \text{Ext}^1(\Omega^1(M),\Lambda) = \text{Ext}^2_\Lambda(M,\Lambda)$ .  The result now follows by

induction.

(f)  (i) and (ii) follow from previous results by means of standard

arguments.  (iii)  follows from the fact that  $\text{Ext}^1_\Lambda(M,\Lambda) = 0$ .  (iv)

follows from previous parts of (f).

(g)  follows from the fact that  $\text{Hom}_\Lambda( \ ,\Lambda) \cong \text{Hom}_T( \ ,T)$  and

$\text{Ext}^1_\Lambda(M,\Lambda) = 0$  for  M  in  $\underline{latt}(\Lambda)$ .

In order to state the main result of this section it is convenient

to have the following definitions.

Suppose  $\Lambda$  is a Gorenstein R-algebra.  Let  M  be a module in

$\underline{latt}(\Lambda)$  which has no nonzero projective summands.  Suppose that

$$\cdots \xrightarrow{f_{i+1}} P_i \longrightarrow \cdots \xrightarrow{f_1} P_0 \longrightarrow M \longrightarrow 0 \quad \text{is a minimal}$$

projective resolution of  M .  Then we define  $\Omega^0(M) = M$  and  $\Omega^i(M) =$

$\text{Im}(f_i)$  for all  $i > 0$ .  Also let

$$\cdots \xrightarrow{\;g_{i+1}\;} Q_i \xrightarrow{\hspace{2cm}} \cdots \xrightarrow{\;g_1\;} Q_0 \xrightarrow{\hspace{1.5cm}} \mathrm{Hom}_\Lambda(M,\Lambda) \xrightarrow{\hspace{1.5cm}} 0 \quad \text{be a}$$

$\Lambda^{op}$ minimal projective resolution for the $\Lambda^{op}$-module $\mathrm{Hom}_\Lambda(M,\Lambda)$ in

$\underline{\mathrm{latt}}(\Lambda)$ which also has no projective summands. Then

$$0 \to \mathrm{Hom}_{\Lambda^{op}}(\mathrm{Hom}_\Lambda(M,\Lambda,\Lambda^{op}) \to \mathrm{Hom}_{\Lambda^{op}}(Q_0,\Lambda^{op}) \xrightarrow{\;\mathrm{Hom}_{\Lambda^{op}}(g_1,\Lambda^{op})\;} \cdots \cdots \cdots \cdots \to \mathrm{Hom}(Q_i,\Lambda^{op}) \to$$

$$\xrightarrow{\;\mathrm{Hom}_{\Lambda^{op}}(g_{i+1},\Lambda^{op})\;} \cdots \cdots \cdots \cdots$$
is an exact sequence of $\Lambda$-module with $M =$

$\mathrm{Hom}_{\Lambda^{op}}(\mathrm{Hom}_\Lambda(M,\Lambda),\Lambda)^{op}$ and the $\mathrm{Hom}_{\Lambda^{op}}(Q_i,\Lambda^{op})$ projective $\Lambda$-modules.

Define $\Omega^{-i}(M) = I_m(\mathrm{Hom}_{\Lambda^{op}}(g_i,\Lambda^{op}))$ for all $i \geq 1$. It is easily

verified that $\Omega^j(\Omega^k(M)) = \Omega^{j+k}(M)$ for all $j$ and $k$ in $Z$, the group

of integers.

<u>Theorem 6.5.</u> Let $\Lambda$ be a Gorenstein R-algebra with $d = \dim(R)$.

(a)  If $A$ is a nonprojective indecomposable $\Lambda$-lattice, then

there is a $\underline{\mathrm{latt}}(\Lambda)$ almost split sequence $0 \to A \to B \to C \to 0$.

(b)  If $C$ is a nonprojective indecomposable $\Lambda$-lattice then there

is a $\underline{\mathrm{latt}}(\Lambda)$ almost split sequence $0 \to A \to B \to C \to 0$.

(c)  If $0 \to A \to B \to C \to 0$ is a $\underline{\mathrm{latt}}(\Lambda)$ almost split sequence,

then,

(i)  $A = \Omega^{2-d}C$,

(ii)  $\Omega^{d-2}A = C$.

<u>Proof:</u>  (a)  Since $\Lambda$ is Gorenstein, we know that $\mathrm{Hom}_T(A,T) = \mathrm{Hom}_\Lambda(A,\Lambda)$.

Also we know that $\mathrm{Hom}_\Lambda(A,\Lambda)$ is not $\Lambda^{op}$-projective since $A$ is not

projective. Thus the fact that there is a $\underline{\mathrm{latt}}(\Lambda)$ almost split sequence $0 \to A \to B \to C \to 0$ follows from Theorem 5.3 because $\Lambda$ is a $\Lambda$-lattice since $\Lambda$ is Gorenstein.

(b) Again a consequence of Theorem 5.3.

(c) (i) If $0 \to A \to B \to C \to 0$ is a $\underline{\text{latt}}(\Lambda)$ almost split sequence we know by Proposition 5.7 that $A$ is an indecomposable summand of $\text{Hom}_T(K,T)$ where $K$ is defined by the exact sequence of $\Lambda^{op}$-modules $0 \to K \to P_{d-1} \to \cdots \to P_0 \to \text{Tr}(C) \to 0$ , where $P_{d-1} \to \cdots \to P_0 \to \text{Tr}(C) \to 0$ is the beginning of a minimal projective resolution for $d \geq 1$ and $K = \text{Tr}(C)$ for $d = 0$ . Since $\Lambda$ is a Gorenstein R-algebra we know that $\text{Tr}(C)$ is an indecomposable $\Lambda^{op}$-lattice, so $K$ is indecomposable and hence $A = \text{Hom}_T(K,T) = \text{Hom}_\Lambda(K,\Lambda)$ . Since $\text{Tr}(C)$ is in $\underline{\text{latt}}(\Lambda)$ , $0 \to \text{Hom}_{\Lambda^{op}}(\text{Tr}(C),\Lambda^{op}) \to \text{Hom}_{\Lambda^{op}}(P_0,\Lambda^{op}) \to \cdots \to \text{Hom}_{\Lambda^{op}}(P_{d-1},\Lambda^{op}) \to \text{Hom}_\Lambda(K,\Lambda) \to 0$ is exact and so $A = \text{Hom}_\Lambda(K,\Lambda) = \Omega^0(\text{Hom}_{\Lambda^{op}}(\text{Tr}(C),\Lambda^{op}))$ . By Proposition 6.4, $\text{Hom}_{\Lambda^{op}}(\text{Tr}(C),\Lambda^{op}) = \Omega^2(C)$ , thus $A = \Omega^{2-d}(C)$ .

(ii) follows from (i) and properties of the operation $\Omega$ .

As an immediate consequence of this theorem we have the next corollary.

**Corollary 6.6.** Let $\Lambda$ be a Gorenstein R-algebra with $\dim(R) = d \geq 2$ . Suppose $A$ is an indecomposable nonprojective $\Lambda$-lattice. Then $\text{Ext}_\Lambda^{d-1}(A,A) \neq 0$ .

Proof: By the previous theorem we know there is a $\underline{\text{latt}}(\Lambda)$ almost split sequence $0 \to A \to B \to \Omega^{d-2}A \to 0$ which is not a split sequence when

$d \geq 2$ . Hence, $\mathrm{Ext}^1(\Omega^{d-2}A,A) \neq 0$ when $d = 2$ , $\Omega^{d-2}A = \Omega^0 A$ and

we have our result. If $d > 2$ , then we have an exact sequence $0 \rightarrow$

$\Omega^{d-2}A \rightarrow P_{d-3} \rightarrow \cdots \rightarrow P_0 \rightarrow A \rightarrow 0$ with the $P_i$ projective modules.

Hence, $\mathrm{Ext}_\Lambda^1(\Omega^{d-2}A,A) \cong \mathrm{Ext}_\Lambda^{d-2+1}(A,A)$ and so $\mathrm{Ext}_\Lambda^{d-1}(A,A) \neq 0$ .

   As a consequence of this corollary we obtain the following theorem.


Theorem 6.7.  Let  T  be a Gorenstein local ring (not necessarily

complete) of dimension $d \geq 2$ .  Suppose $\Lambda$ is an R-algebra which is a

Cohen-Macauley module such that $\Lambda \cong \mathrm{Hom}_T(\Lambda,T)$ as a two-sided $\Lambda$-

module.  If  M  is a $\Lambda$-module satisfying

   (a)  M  is in  $\underline{\underline{\mathrm{CM}}}(\Lambda)$ ,

   (b)  $M_{\underline{p}}$ is a $\Lambda_{\underline{p}}$-projective for prime ideals $\underline{p}$ of  T  such that

        $\dim(T_{\underline{p}}) \leq 1$ , and

   (c)  $\mathrm{Ext}_\Lambda^i(M,M) = 0$ for $i = 1,\ldots,d - 1$ ,

then  M  is a projective $\Lambda$-module.

## REFERENCES

1.  Auslander, M.  "Large modules over Artin algebras," <u>Algebra, Topology</u> <u>and Categories</u>, Academic Press, New York, 1976.

2.  _____.  "Representation theory of Artin algebras, I," <u>Comm.</u> <u>in Algebra</u> (1974), 177-268.

3.  _____.  "Representation theory of Artin algebras, II," <u>Comm.</u> <u>in Algebra</u> (1974), 269-310.

4.  Auslander, M. and I. Reiten.  "Representation theory of Artin algebras, III," <u>Comm. in Algebra</u> (1975), 239-294.

5.  _____.  "Representation theory of Artin algebras, IV," <u>Comm. in Algebra</u>, to appear.

6.  _____.  "Stable equivalence of dualizing R-varieties," <u>Advances in Math.</u> (1974), 300-366.

7.  Bourbaki, N.  <u>Algèbra Commutative</u>, Chapitre 4, Hermann, Paris, 1961.

8.  Herzog, J. and E. Kunz.  <u>Der kanonische Modul eines Cohen-Macauley</u> <u>Rings</u>, Springer-Verlag, New York, 1971.

# NONCOMMUTATIVE PURELY INSEPARABLE EXTENSIONS

Moss Swee3dler

Cornell University

§1.  **Introduction.**  I shall give the definition and survey some results
of general purely inseparable ring extensions.  The notion of pure inseparability presented here generalizes the notion of purely inseparable
field extension in much the same way that the Auslander-Goldman definition of separability generalized the notion of separable field extension.
When  A  is a purely inseparable ring extension of  B  as defined in
Section Two and  B  is central in  A  then  A  may be called a purely
inseparable B-algebra.  This definition of "purely inseparable algebra"
differs from Holleman's in [3], although the two definitions agree when
B  is a field.  Holleman's concept of inseparability may behave better
functorially than the definition I present.

   The second section contains the definition of "purely inseparable"
and a comparison of this definition with the Auslander-Goldman definition
of separable.  The third section goes into properties and characterizations of purely inseparable algebras.  The fourth section shows how the
notion of purely inseparable extension may be used to characterize
Wedderburn factors and potential Wedderburn factors — i.e., maximal

separable subalgebras which become Wedderburn factors on extending the
base field to the algebraic closure.

## §2. Definition of Purely Inseparable Extension.

Suppose $A$ is a ring with subring $B$. (All rings have unit, subrings have the same unit as the full ring.) The **opposite ring** to $A$ is denoted $A^O$ and has elements $\{a^O\}_{a \in A}$. $A \xleftrightarrow{\ a \longleftrightarrow a^O\ } A^O$ is a multiplication reversing correspondence. $A^O$ is a left B-module by $b \cdot a^O = (ba)^O$. Forming $A \otimes_B A^O$ — the tensor product with slip-by $ab \otimes a^O = a \otimes (ba)^O$ — the map $\pi_\beta : A \otimes_B A^O \to A$, $a \otimes a^O \to a\alpha$ is well defined.

Suppose $C$ is a commutative ring and $C \xrightarrow{\ \gamma\ } B$ a ring map with image in the center of $A$. Then both $B$ and $A$ are C-algebras and $A^O$ is considered a C-algebra via $C \to A^O$, $c \to \gamma(c)^O$. Form the C-algebra $A \otimes_C A^O$. View $A$ as a left $A \otimes_C A^O$-module via $(a \otimes a^O) \cdot (a' \otimes a'^O) = aa' \otimes (\alpha'\alpha)^O$. Then $\pi_B : A \otimes_B A^O \to A$ is an $A \otimes_C A^O$-module map. The following proposition is easily verified.

**Proposition.** The $\ker \pi_B$ is a small $A \otimes_C A^O$-submodule of $A \otimes_B A^O$ if and only if $\ker \pi_B$ is a small $A \otimes_{C'} A^O$-submodule of $A \otimes_B A^O$ for any other commutative ring $C'$ and ring map $C' \to B$ having image in the center of $A$.

(If $M$ is a left R-module, a submodule $N$ is **small** if for all proper submodules $P$ of $M$, $P + N$ is a proper submodule of $M$.)

Thus the definition below of purely inseparable extension is independent of C .

A is a **purely inseparable extension** of B if ker $\pi_B$ is a small $A \otimes_C A^o$-submodule of $A \otimes_B A^o$ .

This definition of purely inseparable is equivalent to the definition in [4, p.355].

Before surveying results about general purely inseparable extensions I wish to compare the definition above with the definition of separable extension. Hopefully this will show why $\pi_B$ having small kernel is an appropriate definition of pure inseparability.

Suppose A is a C-algebra. View A as an $A \otimes_C A^o$-module as above. In [1] A was defined as a **separable** C-algebra if A is a projective $A \otimes_C A^o$-module. In this case the natural map $\pi_C : A \otimes_C A^o \to A$ splits as an $A \otimes_C A^o$-module map. Furthermore since $A \otimes_C A^o$ is a free rank 1 $A \otimes_C A^o$-module if $\pi_C$ splits then A is a projective $A \otimes_C A^o$-module. Thus A being a separable C-algebra is equivalent to the splitting of $\pi_C$ or ker $\pi_C$ being a direct summand as $A \otimes_C A^o$-modules.

Now suppose A has subring B and C is a commutative ring which maps to B with image lying in the center of A ; so that B , A and $A^o$ are C-algebras. The following is easily verified.

**Proposition.** The map $\sigma : A \to A \otimes_B A^o$ with $\pi_B \sigma = I$ is an $A \otimes_C A^o$-module map if and only if $\sigma$ is an $A \otimes_{C'} A^o$-module map for any other commutative ring C' and ring map $C' \to B$ with image in the center of A .

Thus ker $\pi_B$ is an $A \otimes_C A^O$-module direct summand if and only if
ker $\pi_B$ is an $A \otimes_{C'} A^O$-module direct summand and the following definition of separable extension is independent of $C$.

The C-algebra $A$ is a **separable** **extension** of $B$ if ker $\pi_B$ is
an $A \otimes_C A^O$-module direct summand of $A \otimes_B A^O$.

Although I do not know a specific reference for this definition, I
believe it appears in the literature.  Perhaps in the work of Hattori.
This definition of separable extension prompts the riddle:

"**Separable** is to **purely** **inseparable** as  (ker $\pi_B$ is a)  **direct**
**summand** is to ??? "

And "**small submodule**" seems a reasonable answer.

§**3.** **Results about Purely Inseparable Algebras.**  Results in this section
are due either to Bogart or myself and most of them can be found in [2]
or [4].

Throughout this section we are considering a ring $A$ with central
subfield $K$.

If $A$ happens to be a field and a purely inseparable K-algebra then
$A$ is a purely inseparable field extension of $K$ in the ordinary field
theoretic sense.  So we will not get in any trouble with two meanings of
"purely inseparable."  This field extension result does not require $A$
to be finite dimensional over $K$.  Perhaps this is a good point to
remind you that an infinite dimensional separable field extension is

not a separable algebra according to the Auslander-Goldman definition.

If  A  is a division ring and purely inseparable K-algebra then  A
is commutative and consequently a purely inseparable field extension of
K .

Now merely suppose that  A  is a purely inseparable K-algebra.  Then
A  is algebraic over  K  , in fact, every element of  A  satisfies a
purely inseparable polynomial in  K[X] ; purely inseparable polynomials
being the polynomials  $\{(X - \lambda)^n\}_{n=1, \lambda \in K}^{\infty}$ .  Since  A  is algebraic over
K  the Jacobson radical — J(A) — is nil.  Moreover,  A/J(A)  is a
purely inseparable field extension of  K .

It is an open question whether  J(A)  being nil and  A/J(A)  being
a purely inseparable field extension of  K  is equivalent to  A  being
a purely inseparable K-algebra.  However,  J(A)  being nil and  A/J(A)
being a purely inseparable field extension of  K  is equivalent to every
element of  A  satisfying a purely inseparable polynomial in  K[X] .

Here are two last results characterizing purely inseparable algebras
before moving on to general purely inseparable extensions.


Proposition.  A  is a purely inseparable K-algebra if and only if
$(A \otimes_K A^o)/J(A \otimes_K A^o)$  is a division ring; in which case, it is a purely
inseparable field extension of  K  isomorphic to  A/J(A) .



Proposition.  If  A  is a finite dimensional K-algebra then  A  is a
purely inseparable K-algebra if and only if the only separable  K

subalgebra of  A  is  K  itself; i.e., if  $A \supset B \supsetneq K$  and  B  is a
K-subalgebra then  B  is not a separable K-algebra.  In this case, all
subalgebras of  A  are actually purely inseparable K-algebras.

§4.  <u>Results about Purely Inseparable Extensions</u>.  The results in this
section are due to Bogart and can be found in [2].

   Now that there is a definition of separable and purely inseparable
one wonders how well the result from field theory <u>that an algebraic
field extension is a separable extension topped by a purely inseparable
extension</u> carries over.  The setting throughout this section is:  A  is
a ring with central subfield  K  and  A  has finite K-dimension.

   When  A  is commutative then  A  has a unique maximal separable
subalgebra  B  and  A  is purely inseparable extension of  B .  In
general, when  A  is not commutative  A  need not have a <u>unique</u> maximal
separable subalgebra, the maximal separable subalgebras of  A  need not
be isomorphic nor of the same dimension and  A  may not be a purely
inseparable extension of a maximal separable subalgebra.  The maximal
separable subalgebras over which  A  is purely inseparable are the
good ones.  This will become clear as we go into Wedderburn Theory.

   A <u>Wedderburn factor</u> for  A  is a separable subalgebra  B  with
$A = J(A) \oplus B$ .

   A <u>Wedderburn spector</u> for  A  is a separable subalgebra  B  with

$$A \otimes_K \bar{K} = J(A \otimes_K \bar{K}) \oplus (B \otimes_K \bar{K}) .$$

Let $J(\ )$ indicate the Jacobson radical and $\bar{K}$ indicate the algebraic closure of $K$. Actually $\bar{K}$ may be replaced by any field extension $L \supset K$ where $(A \otimes_K L)/J(A \otimes_K L)$ is a separable L-algebra.

If $L'$ is any field extension of $K$ then $J(A \otimes_K L') \supset J(A) \otimes_K L'$ and the containment is proper in general, i.e., the radical "grows" under field extension. This "growth" stops once $(A \otimes_K L')/J(A \otimes_K L')$ becomes a separable L'-algebra. Put slightly differently:

**Proposition.** $J(A \otimes_K L') = J(A) \otimes_K L'$ for all field extensions $L' \supset K$ if and only if $A/J(A)$ is a separable K-algebra.

The Wedderburn Principal Theorem tells us that when $A/J(A)$ is a separable K-algebra, there is a separable subalgebra $B$ of $A$ with $A = J(A) \oplus B$. When $A/J(A)$ is not separable — so that $J(A)$ is not done "growing" — there might be a separable subalgebra $B$ of $A$ which is the proper candidate for

(*) $$A = J(A) \oplus B,$$

but the equality (*) is prevented by $J(A)$ not being "full grown." Such a $B$ is a Wedderburn spector.

Both Wedderburn factors and spectors are **maximal** separable sub-algebras of $A$. If $A/J(A)$ is separable then any Wedderburn spector for $A$ is a Wedderburn factor. $A/J(A)$ is separable if $A$ has a Wedderburn factor since then $A/J(A)$ is isomorphic to the Wedderburn factor, Thus if $A$ has a Wedderburn factor, then all Wedderburn spectors are Wedderburn

factors.  In fact, if  A  has a Wedderburn factor then all maximal
separable subalgebras are Wedderburn factors.

Wedderburn factors are always Wedderburn spectors.

I do not wish to give the impression that when  J(A)  is not "full
grown" — i.e.,  A/J(A)  is not separable — there is always a Wedderburn spector.  The maximal separable subalgebras of  A  may not be
"full grown" either.  This can happen in two ways:

(I)    There are algebras where none of the maximal separable
        subalgebras are Wedderburn spectors.

(II)   There are algebras where some of the maximal separable
        subalgebras are Wedderburn spectors and others are not
        Wedderburn spectors.

Bogart obtained the following fine characterization of Wedderburn
spectors.


**Theorem**.  A separable subalgebra  B  of  A  is a Wedderburn spector
if and only if  A  is a purely inseparable extension of  B .


This and other results about spectors are to be found in [2].

Now back to the question of whether an algebra is always a
separable topped by a purely inseparable.  By the theorem, such a
separable subalgebra is a Wedderburn spector.  As mentioned, there
are algebras with no Wedderburn spectors.  So  A  is not always a
separable topped by a purely inseparable.

Maximal separable subalgebras are not unique even when A has Wedderburn factors. Although in this case they are conjugate by inner automorphism. Bogart has given an example where A has non-isomorphic Wedderburn spectors. And in her example of an algebra with some maximal separable subalgebras which are Wedderburn spectors and other maximal subalgebras which are not Wedderburn spectors the two different types of maximal separable subalgebras have different K-dimension. This follows since a separable subalgebra B of A is a Wedderburn spector if and only if the K-dimension of B equals the $\bar{K}$-dimension of $(A \otimes_K \bar{K})/J(A \otimes_K \bar{K})$ . (Again $\bar{K}$ may be replaced by any field extension L of K where $(A \otimes_K L)/J(A \otimes_K L)$ is a separable L-algebra.)

## REFERENCES

1.  Auslander, M. and O. Goldman.  "The Brauer group of a commutative ring," Trans. A.M.S. 97(1960), 367-409.

2.  Bogart, J.  "Wedderburn spectors," to appear.

3.  Holleman, A.  "Inseparable algebras," to appear.

4.  Sweedler, M.  "Purely inseparable algebras," J. of Algebra 35 (1975), 342-355.

# COMMUTATIVE DESCENT THEORY

Moss Swee3dler

Cornell University

§1. **Introduction**. Throughout $A$ is a commutative ring. Suppose $B$ is an A-algebra and $V$ a right B-module. Roughly speaking the **descent problem** is to find a right A-module $X$ where $V \cong X \otimes_A B$ as **B-modules**. When $B$ happens to be a commutative A-algebra and faithfully flat as an A-module the **descent problem** is answered by the theory of faithfully flat descent. When $B = \mathrm{End}_A P$ for $P$ a finitely generated projective A-module the **descent problem** is answered by a special case of Morita descent theory. For a moment forget that $B$ has a ring structure and merely consider $B$ as an A-module. Then $V$ can only be considered as an A-module. We can still look for a right A-module $X$ where $V \cong X \otimes_A B$ as **A-modules**. This problem is answered (in part) by the theory of **twistule descent**. The theory of twistule descent also treats cases where $V$ and $B$ have additional structure, such as $B$ being an A-algebra, $V$ being a B-module and the isomorphism $V \cong X \otimes_A B$ should be a B-module isomorphism. Thus the twistule descent theory gives a common setting for faithfully flat descent and the special case of Morita descent.

The remainder of this article is divided into basically two
sections. The first section (§2) is pure twistule theory. The theory
is discussed from an abstract functorial point of view. A specific
type of descent functor  F  is defined. When such a functor  F  exists
descent is possible. The second section, (§3) through (§6), consists of
applications, including faithfully flat descent, the special case of
Morita descent, a (new) **faithful** descent theory and (new) descent
theories where we can descend modules (comodules) for Lie algebras
(co-algebras) over  A . In a sense the functorial theory of the first
section plays the role of **limits** in the freshman calculus. **Limits** make
it all rigorous, but for many applications you need not know about **limits**.

§**2.** **Functorial Twistule Theory.** Suppose  M  is an A-module and  V  is
an A-module of the form  $X \otimes_A M$  where  X  is another A-module. The
twist map is given by  $T : M \otimes_A M \to M \otimes_A M$ ,  $m_1 \otimes m_2 \to m_2 \otimes m_1$ . Then
corresponding to

$$I \otimes T : X \otimes_A M \otimes_A M \to X \otimes_A M \otimes_A M$$

there is

$$\theta : V \otimes_A M \to V \otimes_A M$$

making the diagram commute

$$X \otimes_A M \otimes_A M \xrightarrow{\quad I \otimes T \quad} X \otimes_A M \otimes_A M$$

(*)

$$V \otimes_A M \xrightarrow{\quad \theta \quad} V \otimes_A M \quad .$$

An easy computation with transpositions shows that, as an endomorphism

of $X \otimes_A M \otimes_A M \otimes_A M$, $I \otimes T \otimes I$ satisfies,

$(I \otimes T \otimes I)(I \otimes I \otimes T)(I \otimes T \otimes I) = (I \otimes I \otimes T)(I \otimes T \otimes I)(I \otimes I \otimes T)$ .

This implies that, as an endomorphism of $V \otimes_A M \otimes_A M$, $\theta \otimes I$ satisfies

(**)        $(\theta \otimes I)(I \otimes T)(\theta \otimes I) = (I \otimes T)(\theta \otimes I)(I \otimes T)$ .

The name of the game is to start with A-modules $V$ and $M$ and an

A-module map $\theta : V \otimes_A M \to V \otimes_A M$ satisfying (**) and then try to recover

$X$ and recover the isomorphism $V \cong X \otimes_A M$ which makes the diagram (*)

commute. It is not quite so simple. For example, the zero map from

$V \otimes_A M$ to $V \otimes_A M$ satisfies (**). Yet the zero map does not in general

arise from an $I \otimes T$ .

The way to proceed is to fix $M$ and call a pair $(V, \theta)$ an __M-twistule__

if (**) is satisfied.

The M-twistule is called __vert__ if $\theta$ is a bijection. The category

of vert M-twistules is denoted $T_M$ and the category of A-modules is

denoted $M_A$ . The main question is when the functor $S$ given by

$$S : M_A \rightsquigarrow T_M$$
$$X \longrightarrow (X \otimes_A M, I \otimes T)$$

is an equivalence of categories. A functor $F : T_M \rightsquigarrow M_A$ satisfying three conditions is called a **descent functor**. The first condition on a descent functor is that the composite functor,

(I)                                  $M_A \overset{S}{\rightsquigarrow} T_M \overset{F}{\rightsquigarrow} M_A$

is naturally equivalent to the identity functor of $M_A$ . We must introduce some other functors in order to present the remaining two conditions for $F$ to be a descent functor.

For any A-module $Y$ the functor $M_A \rightsquigarrow M_A$, $X \rightarrow Y \otimes_A X$ is denoted $\boxed{Y\otimes}$ . If $(V,\theta)$ is a vert M-twistule then $(Y \otimes_A V, I \otimes \theta)$ is also a vert M-twistule. The functor $T_M \rightsquigarrow T_M$ , $(V,\theta) \rightarrow (Y \otimes_A V, I \otimes \theta)$ is also denoted $\boxed{Y\otimes}$ . The second condition on a descent functor $F$ is that:

(II)                  $F\boxed{M\otimes}$  is naturally equivalent to  $\boxed{M\otimes}F$ .

(Not for all $\boxed{Y\otimes}$ , just $\boxed{M\otimes}$ !)

The third condition on a descent functor grows out of the second.

Since $\boxed{M\otimes}$ $\boxed{M\otimes}$ is naturally equivalent to $\boxed{(M \otimes_A M)\otimes}$ the second condition iterated gives a natural equivalence:

$$F\boxed{(M \otimes_A M)\otimes} \overset{\gamma}{\underset{\sim}{\longrightarrow}} \boxed{(M \otimes_A M)\otimes} F .$$

The module map $T : M \otimes_A M \rightarrow M \otimes_A M$ gives a natural transformation

$$T : \boxed{(M \otimes_A M)\otimes} \rightarrow \boxed{(M \otimes_A M)\otimes}$$

where if $U$ is an A-module

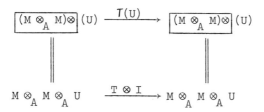

The **third** condition on a descent functor is that the diagram of functors and natural transformations commutes:

(III)

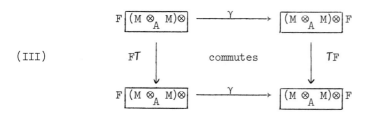

The main descent theorem in twistule theory is:

**Theorem.** If $F : T_M \rightsquigarrow M_A$ satisfies the three conditions above then $F$ or equivalently $S$ is an equivalence of categories. If $(V, \theta) \in T_M$ then $V \cong F(V) \otimes_A M$ as an M-twistule.

When $S$ is an equivalence of categories all vert M-twistules are isomorphic to M-twistules of the form $(X \otimes_A M, I \otimes T)$ . Such M-twistules are called **split** M-twistules. There is a splitting theory for M-twistules which is needed to prove the descent theorem.

§3.  Applications.  An application consists minimally of the following:

(1)  choice of  A ,

(2)  choice of  M , and

(3)  construction of a descent functor

$$F : T_M \to M_A .$$

Then, by the theory of the previous section, each M-twistule is of the

form  $X \otimes_A M$  as an A-module.  Often one wishes to consider additional

structures such as  M  being an algebra, co-algebra or Lie algebra.  In

these cases  $X \otimes_A M$  has the additional structure of M-module, M-comodule

or M-module, respectively.  The additional structure is easily taken

into account.  If  (V,θ)  is an M-twistule and  V  is an M-module (or

M-comodule), then  $V \otimes_A M$  has two M-module (or M-comodule) structures.

The first arising from the module (comodule) structure of  V ; the second

arising from the natural module (comodule) structure of  M .  Since

$θ : V \otimes_A M \to V \otimes_A M$ , it is natural to require that  θ  interchange the

two module (comodule) structures.  That is, if the  $V \otimes_A M$  which is the

domain of  θ  has one of the module (comodule) structures and the

$V \otimes_A M$  which is the range of  θ  has the other, then  θ  should be a

module (comodule) map.  In this case we say that  θ  is an  M  **action map**.

If there is a descent functor  $F : T_M \to M_A$  which satisfies a

slightly strengthened version of (III) in the previous section then

descent respects the additional structure as follows:

**Theorem.** Suppose $(V, \theta) \in T_M$ , V is an M-module (comodule) and $\theta$ is an M **action map**; let $V \overset{\xi}{\cong} F(V) \otimes_A M$ be the isomorphism given by the descent theory; then $\xi$ is an isomorphism of M-modules (comodules).

§**4.** **Faithfully Flat Descent.** When M is an A-algebra let

$E_1 : X \to X \otimes_A M, x \to x \otimes 1$ for all A-modules X . The functor

$F_1 : T_M \leadsto M_A$ is defined by

$$(V, \theta) \to \ker(V \xrightarrow{\quad E_1 - \theta E_1 \quad} V \otimes_A M) \ .$$

If M is a faithfully flat A-module then $F_1$ is a descent functor and satisfies the strengthened version of (III). Thus for $(V, \theta) \in T_M$ where V is an M-module and $\theta$ an M-action map the isomorphism $V \cong F_1(V) \otimes_A M$ is an M-module isomorphism.

This $V \cong F_1(V) \otimes_A M$ is the usual faithfully flat descent correspondence. $\theta$ is not new to the faithfully flat theory either. In the faithfully flat theory $\theta$ is called "descent data." The twistule condition, $(\theta \otimes I)(I \otimes T)(\theta \otimes I) = (I \otimes T)(\theta \otimes I)(I \otimes T)$ is equivalent to $\theta$ being a "cocycle" in the faithfully flat theory.

§**5.** **Faithful Descent, Lie Descent and Co-algebra Descent.** Here M is merely assumed to be an A-module and $e : M \otimes_A N \to A$ a surjective A-module map where N is some other A-module. This implies that $\boxed{M \otimes}$

is a faithful functor but does not imply that  M  is flat.   The functor
$F_2 : T_M \to M_A$  is defined by,

$$(V,\theta) \to \text{coker} \left( \begin{array}{c} V \otimes_A M \otimes_A N \otimes_A N \\ \\ \downarrow (I \otimes e \otimes I)(I \otimes I \otimes T - \theta \otimes I \otimes I) \\ \\ V \otimes_A A \otimes_A N = V \otimes_A N \end{array} \right)$$

$F_2$  is a descent functor and satisfies the strengthened version of (III).
Thus each M-twistule  $(V,\theta) \in T_M$  is isomorphic to  $F_2(V) \otimes_A M$  as M-
twistules.   If  M  happens to be an A-Lie algebra, then the isomorphism
$V \cong F_2(V) \otimes_A M$  is an isomorphism as M-modules when  $\theta$  is an M-action
map.

For co-algebra descent  N  often disappears.   If  M  is an A-co-alge
there is a map  $\varepsilon : M \to A$  which is usually surjective.   Thus  N  can be
chosen as  A  and the picture becomes

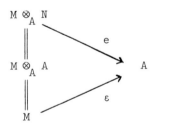

The functor  $F_2$  then reduces to  $(V,\theta) \to \text{coker}\left(V \otimes_A M \dfrac{I \otimes \varepsilon - (I \otimes \varepsilon)\theta}{}\right.$
which is dual to the  $F_1$  from faithfully flat descent.   The co-algebra
descent theory is slightly better than the faithfully flat theory.

$\varepsilon : M \to A$ being surjective implies that $\boxed{M \otimes}$ is faithful but $M$ need not be flat as an $A$-module.

In the co-algebra descent theory $V \cong F_2(V) \otimes_A M$ is an $M$-comodule isomorphism for $M$-comodules $V$ where $\theta$ is an $M$-action map.

## §6. Faithful Descent and Commutative Morita Descent.

Here $M$ is assumed to be an $A$-module and $e : M \otimes_A N \to A$ a surjective $A$-module map.

Then $N \otimes_A M$ has the **product** $\gamma$ where $\gamma$ is determined by the diagram:

$$
\begin{array}{ccc}
(N \otimes_A M) \otimes_A (N \otimes_A M) & \xrightarrow{\gamma} & N \otimes_A M \\
\Big\| & & \Big\| \\
N \otimes_A M \otimes_A N \otimes_A M & \xrightarrow{I \otimes e \otimes I} & N \otimes_A A \otimes_A M
\end{array}
$$

Associativity of $\gamma$ is easily verified. In general, there is no unit and so I shall not speak of $N \otimes_A M$ as an $A$-algebra. However, I shall speak of right and left $N \otimes_A M$-modules. The maps

$$
M \otimes_A (N \otimes_A M) \xrightarrow{e \otimes I} A \otimes_A M = M ,
$$

$$
(N \otimes_A M) \otimes_A N \xrightarrow{I \otimes e} N \otimes_A A = N
$$

give $M$ a right $N \otimes_A M$-module structure and $N$ a left $N \otimes_A M$-module structure, respectively. (If there is $x \in N \otimes_A M$ which acts as the

identity on $M$ then $M$ is a finitely generated projective A-module.
In general, $M$ is neither finitely generated nor flat — hence not
projective.)

The link between faithful descent theory and the special case of
Morita descent lies in the possibility of passing from M-twistules to
$N \otimes_A$ M-modules and vice versa. If $(V,\theta)$ is an M-twistule then the
map

$$V \otimes_A (N \otimes_A M) \xrightarrow{\ I \otimes T\ } V \otimes_A M \otimes_A N \xrightarrow{\ \theta \otimes I\ } V \otimes_A M \otimes_A N$$

$$\xrightarrow{\ I \otimes e\ } V \otimes_A A = V$$

gives $V$ a right $N \otimes_A$ M-module structure. (The twistule condition on
$\theta$ gives the module associativity condition $(v \cdot x) \cdot y = v \cdot (xy)$ .)
The functor $F_2$ from faithful descent (defined above) reduces to

$$(V,\theta) \to V \otimes_{(N \otimes_A M)}{}^N \ ,$$

where tensoring over $N \otimes_A M$ is like tensoring over any ring. The
correspondence $V \to V \otimes_{(N \otimes_A M)}{}^N$ is exactly the Morita descent
correspondence.

The remaining question is how to go from $N \otimes_A$ M-modules to M-
twistules. To do this we must assume that $M$ is a finitely generated
projective A-module, $N = \text{Hom}_A(M,A)$ and

$$e : M \otimes_A N \to A \ , \quad m \otimes f \to f(a) \ .$$

In this case, the right $N \otimes_A M$-module structure on $M$ gives a ring

isomorphism $N \otimes_A M \cong \operatorname{End}_A M$ if we consider the functions in $\operatorname{End}_A M$

acting (written) on the right on $M$ .

Now how do we get from a right $\operatorname{End}_A M$-module to an M-twistule?

Suppose $U$ is a right $\operatorname{End}_A M$-module. Then $U \otimes_A M$ is a right

$(\operatorname{End}_A M) \otimes_A (\operatorname{End}_A M)$-module. Since $M$ is a finitely generated projective

A-module, there is a natural A-algebra isomorphism

$$(\operatorname{End}_A M) \otimes_A (\operatorname{End}_A M) \cong \operatorname{End}_A (M \otimes_A M) .$$

Under this isomorphism let the twist map $T \in \operatorname{End}_A (M \otimes_A M)$ correspond

to $\Omega \in (\operatorname{End}_A M) \otimes_A (\operatorname{End}_A M)$ . Let $\omega : U \otimes_A M \to U \otimes_A M$ be defined by

$z \to z \cdot \Omega$ for $z \in U \otimes_A M$ . Then it can be shown that $(U, \omega)$ is an

M-twistule and this correspondence from right $\operatorname{End}_A M$-modules to M-

twistules is the inverse correspondence to

$$\{\text{M-twistules}\} \longrightarrow \{N \otimes_A \text{ M-modules}\} \longrightarrow \{\operatorname{End}_A \text{M-modules}\} .$$

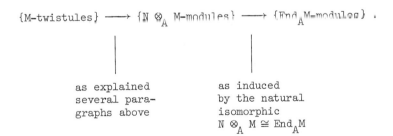

|   as explained | as induced |
|   several para- | by the natural |
|   graphs above | isomorphic |
|   | $N \otimes_A M \cong \operatorname{End}_A M$ |

Thus the faithful descent functor $F_2$ applied to $(U, \omega)$ gives

$U \otimes_{\operatorname{End}_A M} N$ .

Of course, $M$ being a finitely generated projective A-module and

right $\operatorname{End}_A M$-modules $U$ descending to $U \otimes_{\operatorname{End}_A M} \operatorname{Hom}_A (M, A)$ is exactly

the Morita descent theory.  Full Morita descent theory is better since
A  need not be commutative.  The faithful descent theory is better in
a different way.  A  must be commutative but  M  need not be finitely
generated nor flat.

## REFERENCES

1.  Bass, H.  The Morita Theorems, mimeographed notes, The University
    of Oregon, Eugene, Oregon, 1962.


2.  Grothendieck, A.  "Technique de Descente et Théorèmes d'Existence
    en Géométrie Algébrique," I. Sém. Bourbaki #190(1959-1969).


3.  Sweedler, M.  "Twistules and descent," to appear.

BRAUER GROUPS

Daniel Zelinsky

Northwestern University

§1.  **Introduction.**  In these talks, I am under strict orders to give a
survey of the subject for the nonexpert.  To those of you who are
experts, I apologize for the elementary character of much of my
material.  Besides, my survey is far from complete and definitive.
There are many facets of the study of Brauer groups which are only
mentioned here, and many mathematicians whose important contributions
are not included at all.  I can only hope to sketch the history of
these groups and give the flavor of the research that is continuing
today.

§2.  **Classical Brauer Groups.**  I suppose it all began with Hamilton
and his quaternions.  A **quaternion** is the sum of a real scalar and
a vector in three-space.  Addition of quaternions is defined in the
obvious way.  Multiplication is easy:  the scalar part multiplies
scalars and vectors in the expected way; and the product of two
vectors is  $uv = -u \cdot v + u \times v$ , using both the ordinary dot and
cross-products in Euclidean three-space.  Surely Hamilton was intrigued

that this simple artifice not only encapsulates all the ordinary operations with vectors and scalars, but also defines a four-dimensional algebra that is an excellent imitation of the complex numbers:  It contains the reals (and the complex numbers) and it has addition, subtraction, multiplication, division, and absolute values behaving just like the complex numbers, except that the commutative law of multiplication is lacking.  The center of the algebra is the ground field of real numbers.  We say the algebra is **central** over this field. In fact, it is a central "division" algebra.

In the 19-th century there was a great quaternionic vogue; people did quaternionic analytic function theory, integration, and everything else you could do with complex numbers.  But none of this has lasted. The future of Hamilton's quaternions was elsewhere.

Frobenius proved that the quaternions form the _only_ algebra which is a central division algebra over the real numbers.  By this time, other algebras were being considered, too:  matrix algebras, group algebras.  But, following the Hamiltonian mental set, they were called hypercomplex systems, or systems of hypercomplex numbers.  This name persisted until well into the twentieth century.  [The search for more imitations of the complex numbers is another story, but a truncated one: Division algebras over the real numbers, even if you give up the commutative and associative laws, can only be 2- , 4- or 8-dimensional and are just the ones that were already known in the 19-th century.  We have to thank topologists for most of this theory.]

The stage for the true future of quaternions was set by Wedderburn, who did for associative algebras in 1905 and 1907 what Elie Cartan had already done for Lie algebras in 1899.  Wedderburn classified finite dimensional algebras, first over the field of complex numbers, then over an arbitrary field.  The latter, as you know, reduced the whole classification to the problem of listing all the division algebras over the field (plus some uncomfortable compromises with radicals):  Every finite dimensional algebra with zero radical is a finite direct product of simple algebras, each of which must be the algebra of all  n  by  n  matrices with entries in a division algebra. Since all division algebras over the reals and over the complex numbers were known, this solved the problem of listing all algebras over those fields, at least all those with zero radical.  Over fields like the rational numbers, it was a different story.  What kinds of division algebras could exist over a field turned out to be a rather deep property of the field, depending on the number theory or other arithmetic properties of the field.

In the late 1920's and early 1930's the classification of all division algebras over the rationals was essentially solved by Hasse, Brauer, and Noether.  We return to this in a moment.  In the process, Brauer noticed that the classification could be neatly summarized and even augmented by making an abelian group of the (isomorphism classes of) central division algebras over a fixed field.  The group operation is tensor product (in those days called "direct product" or "Kronecker product"), except that the tensor product of division algebras is not usually a division algebra; it is one of Wedderburn's simple algebras, a matrix algebra over a division algebra. Throwing away the matrices and keeping only the division algebra of entries, you get a "product" of the two original algebras.

Of course, if we are going to throw away the matrix part of every simple algebra we get as a product, we may as well start with the central simple algebras instead of central division algebras in the first place — they are matrix algebras over division algebras — and identify two of them (call them similar) if their division algebras are isomorphic. In other words, instead of isomorphism classes of central division algebras we use what amounts to the same thing, similarity classes of central simple algebras. These similarity classes  form a monoid under tensor product. The identity is the class of all matrix algebras over the ground field.

[The notion of similarity can be defined as well by adding on matrices instead of throwing them away: A is _similar_ to B if and only if the algebra of m by m matrices with entries in A is isomorphic to the algebra of n by n matrices with entries in B for some m and n .]

Amazing result: Every similarity class has an inverse; the classes form a group. In fact, if A is an algebra in one class, the inverse class is the class of the opposite algebra $A^0$ (the same vector space as A but the product ab in $A^0$ is defined to be the product b times a in A ). The result is amazing, but easy, because if A is an algebra over $F$ then $A$ is embedded in $\text{Hom}_F(A,A)$ by sending

every element of $A$ to left multiplication by that element; and $A^0$

is embedded as right multiplications. This gives an algebra homomorphism

$$A \otimes_F A^0 \to \text{Hom}_F(A,A) .$$

By the preceding remarks, if $A$ is central simple the tensor product

is, too, so the map is a monomorphism. Counting vector space dimensions

over $F$ shows that it is an epimorphism. Thus $A \otimes_F A^0$ is isomorphic

to a matrix algebra $\text{Hom}_F(A,A)$ , so is in the identity class, as

desired.

This abelian group of similarity classes of central simple algebras

over $F$ is called the **Brauer group** of $F$ and is denoted $\text{Br}(F)$ . (For

general reference, see [26] or [2].) As we said, it is a rather deep

invariant of $F$ . It varies sharply with $F$ , and is not known for

very many fields. We list most of the known results.

The Brauer group of any algebraically closed field is zero. The

Brauer group of the field of real numbers is cyclic of order two, ac-

cording to Frobenius. (You can even see how the class of the quaternions

is its own inverse as it must be in this group of order two:  there is

a conjugation in the quaternions:  $a + u \to a - u$ if $a$ is scalar and

$u$ is a vector; this is an anti-automorphism, so an isomorphism of $A$

and $A^0$ .)

The Brauer group of a finite field is zero (Wedderburn's theorem

that all finite division rings are commutative).

Tsen's theorem ([38], extended in [31]) asserts the vanishing of the Brauer group of the field of rational functions of one variable over an algebraically closed field (or of any algebraic extension of such a field).

The Brauer group of the field $Q$ of rational numbers may be described as follows:  The embedding of $Q$ into the field $R$ of real numbers gives a homomorphism $Br(Q) \to Br(R)$ by extending the scalars in each $Q$-algebra $A$, that is by mapping the class of $A$ in $Br(Q)$ to the class of $A \otimes_Q R$ in $Br(R)$.  Similarly, the embedding of $Q$ into its other completions, $\hat{Q}_p$, the field of p-adic numbers, gives a homomorphism $Br(Q) \to Br(\hat{Q}_p)$; but any one algebra class in $Br(Q)$ maps to the identity class in all but a finite number of $\hat{Q}_p$.  Put all these homomorphisms together to get a homomorphism from $Br(Q)$ to the direct sum $Br(R) \oplus \bigoplus_p Br(\hat{Q}_p)$.  Most of the Hasse-Brauer-Noether Theorem [12] consists in proving that this map is a monomorphism, and that $Br(\hat{Q}_p) = Q/Z$ (we make a few remarks about this in the next section).  Of course, $Br(R)$ is $Z/2Z$, which can be identified with the unique subgroup of order two in $Q/Z$.  We can then define a sum map $Br(R) \oplus \bigoplus_p Br(Q_p) \to Q/Z$ which simply adds the entries in each infinituple.  Finally, $Br(Q)$ is isomorphic (by the monomorphism described above) to the kernel of this sum map.  Or, less symmetrically, $Br(Q)$ is isomorphic to $Z/2Z \oplus \bigoplus Q/Z$, where the direct sum of $Q/Z$'s has countably many summands, one for each rational prime but one.

A similar theorem holds for the ring of algebraic integers in any algebraic number field, replacing $R$ and $\hat{Q}_p$ by all the completions of the field in all possible absolute values.  And, of course,

the results are the same for the other "global fields:" algebraic function fields of one variable over finite fields.

There are very few general results. Perhaps the only one is
this: The Brauer group is always a torsion group. For a field of
characteristic $p \neq 0$ , the theory of p-algebras ([2], Ch.VII)
gives a good description of the p-torsion part; in particular, this
part is divisible, so that $Br(F)$ is p-divisible if $F$ is a field of
characteristic $p$ .

In the next section we shall describe an approach to the Brauer
group which has, in one way or another, been at the root of much of the
progress in this subject.

## §3. Cohomological Description of the Brauer Group. We could make a
case for the proposition that the cohomology of groups started here, in
the theory of the Brauer group, or at least in the theory of central
simple algebras. Long before group cohomology had been defined, it was
known that $Br(F)$ was a union of groups which are now called $H^2(G,U(K))$
It came about this way.

The early search for new division algebras looked for generalizations of the algebra of quaternions. Look at the quaternions as an
algebra over the real number field $F$ containing the complex number
field $K = F + Fi$ ; besides this there is the quaternion $j$ which has
the property $jcj^{-1} = \bar{c}$ for every complex number $c$ (here $\bar{c}$ denotes
the conjugate complex number); and $j^2 = -1$ . Generalize by taking any

field  F  and any Galois extension field  K  whose Galois group is cyclic
generated by an automorphism  σ  (replacing complex conjugation).  Adjoin
a new element  j  to  K  with  $jcj^{-1} = \sigma(c)$  for all  c  in  K .  This
means  $jc = \sigma(c)j$  for all  c .  Hence,  $j^i c = \sigma^i(c)j^i$  and the usual
proof of independence of automorphisms shows that  $1,j,\ldots,j^{n-1}$  must
be left linearly independent over  K , where  n  is  [K:F]  or the order
of  σ .  If  $j^n = \alpha$  in  F, we get a multiplication table for a **cyclic
algebra** which turns out to be central simple if  $\alpha \neq 0$  and a division
algebra if  α  is not a norm of an element of  K .  The heart of the
Hasse-Brauer-Noether Theorem is that every division algebra with center
the rational number field is a cyclic algebra.  But for general fields,
this is not true.  In fact, there are algebras which are not even similar
to any cyclic algebra.

A second generalization is needed.

Instead of a cyclic field, use any Galois extension field, with
group  G .  Then we need one  $j_\sigma$  for each  σ  in  G , with  $j_\sigma c j_\sigma^{-1} =$
$\sigma(c)$  for all  c  in  K .  Then, instead of  $j_{\sigma^i} = (j_\sigma)^i$  as in the
cyclic case, and instead of  $j_\sigma^n = \alpha$ , we postulate

$$j_\sigma j_\tau = f(\sigma,\tau)j_{\sigma\tau}$$

for some function  f  from  G × G  to the nonzero elements of  K .  To
make the resulting algebra associative,  f  must satisfy an identity
which currently is known as the 2-cocycle identity:
$\rho(f(\sigma,\tau))f(\rho\sigma,\tau)^{-1}f(\rho,\sigma\tau)f(\rho,\sigma)^{-1} = 1$  for all  ρ,σ,τ  in  G .  Such an
f  is called a **factor set** and automatically defines a central simple F-

algebra.  The algebra is called a **crossed-product** of  K  by  G .

This is general enough.  Every similarity class in the Brauer group
of any field  F  does contain some crossed-product (of some extension
field  K  by its Galois group, with some factor set).  [If we drop from
similarity classes to isomorphism classes, the crossed-products are
not enough.  In 1971, Amitsur showed that not every central simple
algebra is isomorphic to a crossed-product.]

Let us compute one easy crossed-product.  Take any Galois extension  K  of  F  with its Galois group  G , and use the factor set
$f(\sigma,\tau) = 1$  for all  $\sigma,\tau$  in  G .  Then the corresponding crossed-product
is just isomorphic to  $\text{Hom}_F(K,K)$ , a matrix algebra, belonging in the
unit class of  Br(F) .  We can write the isomorphism explicitly.  Each
c  in  K  is mapped to the  endomorphism  $c_r$  of  K , right multiplication by  c .  Each  $j_\sigma$  can be mapped to  $\sigma$ , which is an element of
$\text{Hom}_F(K,K)$ .  This mapping is an algebra homomorphism because
$j_\sigma j_\tau = j_{\sigma\tau}$  when  $f(\sigma,\tau) = 1$ , and because the relation  $j_\sigma c j_\sigma^{-1} = \sigma(c)$
in the crossed-product is matched by the easy identity  $\sigma \circ c_r \circ \sigma^{-1} =$
$\sigma(c)_r$  in the endomorphisms of  K .  A dimension check shows that the
map is an isomorphism.  Therefore the factor set  $f(\sigma,\tau) = 1$  maps to
the identity element of  Br(F) .  But other factor sets do, too.  If we
replace each  $j_\sigma$  in a crossed-product by  $a_\sigma j_\sigma$  where each  $a_\sigma$  is a
nonzero element of  K , we get the same algebra but a new factor set,
the old factor set multiplied by the "trivial factor set"  $t(\sigma,\tau) =$
$\sigma(a_\tau)a_{\sigma\tau}^{-1}a_\sigma$ .  It turns out that the factor sets which give the identity
of  Br(F)  are exactly these trivial ones.  So here is the homology:

Fix  K  and  G  and write  U(K)  for the multiplicative group of units
(nonzero elements) of  K ; the factor sets  $f : G \times G \to U(K)$  are the
cocycles; the trivial factor sets are the coboundaries; and the quotient
group  $H^2(G,U(K))$  maps to  Br(F) .  The map (cocycle → factor set →
crossed-product algebra) is in fact a homomorphism and even a monomorphism.

The image is the subgroup of  Br(F)  consisting of all algebra
classes that may be represented by crossed-products of  K  by  G .  It
is a subgroup of  Br(F)  depending on the Galois extension  K  of  F ,
and is denoted  Br(K/F) .  However, this description of it is not
altogether satisfying.  You can detect a crossed-product of  K  by  G ;
it must have  K  as a maximal commutative subalgebra, and this is both
necessary and sufficient.  [In fact, this is the importance of crossed-
products; they link the theory of central simple (very noncommutative)
algebras with the theory of commutative extensions like  K .]  But to
see whether the class of an algebra  A  is in  Br(K/F)  you would have
to search among all the algebras similar to  A , looking for one which
does contain a maximal commutative subalgebra isomorphic to  K .  It
would be more convenient if there were a criterion depending on  A  itself.  There is a pretty theorem that does it.

__Theorem__.  Br(K/F)  is the kernel of the homomorphism  Br(F) → Br(K)
referred to in Section 2, produced by extending the scalars,  {A} →
$\{A \otimes_F K\}$ .

In concrete terms, the theorem says that a central simple F-algebra  $A$  is similar to a crossed-product of  $K$  by its Galois group  $G$  if and only if the K-algebra  $A \otimes_F K$  is a matrix algebra over  $K$  (we say  $A$  is split by  $K$ ).

In one direction, this theorem is relatively easy:  If  $A$  is a crossed-product of  $K$  by  $G$ , then  $A$  contains  $K$ .  The centralizer of  $K$  in  $A$ , namely,  $\{a \in A \mid ak = ka$  for all  $k$  in  $K\}$ , is again  $K$ .  This implies that  $A \otimes A^0$  contains a copy of  $K$ , namely,  $F \otimes K$ , whose centralizer is  $A \otimes K$ .  Now use the isomorphism  $A \otimes A^0 \to \mathrm{Hom}_F(A,A)$  adduced in Section 2 for all central simple  $A$ .  The copy of  $K$  goes over to the set of right multiplications (on  $A$ ) by elements of  $K$ , and the centralizer of this set is exactly  $\mathrm{Hom}_K(A,A)$  where  $A$  is treated as a right K-module.  Hence,  $A \otimes K \cong \mathrm{Hom}_K(A,A)$ .  This shows that  $A$  is split by  $K$ .  (The argument did not need the fact that  $A$  is a crossed-product of  $K$ , nor even that  $K$  is Galois, but only that  $K$  is a maximal commutative subalgebra of  $A$ , that is,  $K$  is its own centralizer in  $A$ .)  If  $A$  is only similar to an algebra with  $K$  as maximal commutative subalgebra, then  $A$  is also split by  $K$ .  We shall not go into the converse.  The argument we have given points up the main idea:  There are two relations between the commutative  $K$  and the non-commutative crossed-product  $A$  of  $K$  by  $G$ :  $K$  is a maximal commutative subalgebra of  $A$ , and  $K$  splits  $A$ .  The latter property is shared by all algebras in the same similarity class as  $A$ , while the former is not.  Furthermore, the latter is defined even for non-Galois  $K$ .  And, finally, the splitting condition generalizes better in the

next section. So, from now on, we adopt it; $Br(K/F)$ will mean the
subgroup of $Br(F)$ consisting of classes of algebras split by $K$ .

In any case, the $Br(K/F)$'s with $K$ ranging over the Galois
extensions of $F$ , form a collection of subgroups of $Br(F)$ . In fact,
their union is all of $Br(F)$ ; that is, every central simple F-algebra
is $A$ is split by some Galois extension of $F$ : It takes little more
than Zorn's Lemma to produce a maximal commutative subfield of $A$ .
A tricky computation produces a separable one. The argument just
given shows that $A$ is split by this separable field. But then it
is split by any larger field, in particular, by some Galois extension field.

We have the first homological description of the Brauer group of
a field $F$ : it is the union of $H^2(G,U(K))$ as $K$ ranges over the
Galois extensions of $F$ ; or it is $H^2(\bar{G},U(\bar{K}))$ where $\bar{G}$ is the
compact (profinite) Galois group of the separable closure $\bar{K}$ of $F$ ,
and the cohomology is the continuous cohomology. Except for the
homological terminology, all this dates back to the 1920's and earlier.

More recently, Amitsur introduced a complex in [3], showing that even
if $K$ is not a Galois extension, $Br(K/F)$ can be described as a 2-co-
homology group. His complex is extremely natural: Consider the F-alge-
bras $K$ , $K^2 = K \otimes_F K$ , $K^3 = K \otimes_F K \otimes_F K$ , ... . There are $n + 1$ standard algebra homomorphisms from each $K^n$ to the next $K^{n+1}$ . For example,
$K^2 \to K^3$ by sending $x \otimes y$ either to $1 \otimes x \otimes y$ , or to $x \otimes 1 \otimes y$ , or t
$x \otimes y \otimes 1$ . These algebra homomorphisms induce group homomorphisms
called **face maps** from the group of units $U(K^n)$ to $U(K^{n+1})$ . The

alternating product of these  n + 1  face maps is a single, boundary
map       $U(K^n) \to U(K^{n+1})$ .  Do this for every  n  and we get the
Amitsur complex

$$U(K) \to U(K^2) \to U(K^3) \to \cdots .$$

The homology at  $U(K^{n+1})$  is denoted  $H^n(K/F,U)$ .  Amitsur proved that
$H^2(K/F,U)$  is isomorphic to  $Br(K/F)$  for every finite field extension
K  of  F .

    If we examine the special case of a Galois extension  K  with Galois
group  G , the algebra  $K^2$  is just a direct sum of copies of  K , or,
better, the set of all functions from  G  to  K :  Given  σ  in  G  and
$\Sigma \, x_i \otimes y_i$  in  $K^2$  we have an element of  K , namely,  $\Sigma \, x_i \sigma(y_i)$ .
Fixing the element of  $K^2$  and letting  σ  vary, we associate to each
element of  $K^2$  a function from  G  to  K .  Routine checks then show
that this is a bijection, and that, similarly,  $K^n$  (to  n  factors)
is isomorphic to the ring of all functions from  $G \times \cdots \times G$  to  K .
A function is invertible if and only if all the functional values are
invertible (i.e., in  U(K) ) so  $U(K^n)$  is isomorphic to the multiplicative group of all functions from  $G \times G \times \cdots \times G$  to  U(K) .  This
is the group of (homogeneous) cochains in the cohomology theory.
Further cacluations show that this isomorphism carries Amitsur's
"face maps" into the ordinary face maps used in making up the coboundary operator in the cohomology of  G , so Amitsur's cohomology
in this case is the same as group cohomology.

In summary, $Br(F)$ is the union of subgroups $Br(K/F)$ . Each
$Br(K/F)$ is the kernel of the scalar extension homomorphism $Br(F) \to Br(K)$
and is isomorphic to an Amitsur cohomology group $H^2(K/F,U)$ , or, if $K$
is Galois over $F$ with group $G$ , to the group cohomology $H^2(G,U(K))$ .

As an example of the use of this homological description, we
can prove that $Br(F)$ is a torsion group. Since $Br(F)$ is the union
of all the $Br(K/F)$ , it suffices to prove the latter is a torsion group.
But it is isomorphic to $H^2(G,U(K))$ and any cohomology group of the
finite group $G$ is annihilated by the order of $G$ .

The homological method is also exploited to prove $Br(\hat{Q}_p) = Q/Z$
[37]. Very roughly, the close connection between $\hat{Q}_p$ and its residue
field $Z/pZ$ allows every central simple $\hat{Q}_p$-algebra to be split by a
cyclic field extension $K$ determined by an extension of $Z/pZ$ . And
for such a $K$ , $H^2(G,U(K))$ can be computed quite explicitly; it
turns out to be a cyclic group of order $[K : \hat{Q}_p]$ and there is one such
extension of each dimension. It is not difficult, then, to see $Q/Z$
as the union of these finite cyclic groups.

§4.  <u>Brauer Groups of Rings</u>. An absolutely natural impulse in virtually
all of algebra is to do for commutative rings what has already been
done for fields. However, no such development occurred for Brauer
groups until 1951, when Azumaya wrote, "On maximally central algebras"
[9]. Algebras over rings had been much studied, of course, in connection with number theory (algebras over the ring of integers) and

group theory (group rings over the integers or over other rings).  In
general, the algebras were algebras over integral domains, hence orders
in algebras over fields.  But there is no Wedderburn theory to reduce
the study of algebras over rings to the study of some analog of division
algebras.  And none of the algebras were simple, either, as they were in
the penultimate step of the Wedderburn structure theory.  Nevertheless,
Azumaya found one class of algebras over any commutative ring which
generalized the class of central simple algebras over a field.  These
algebras could be exploited to give some invariants of the ground ring,
although they could not be used as in the Wedderburn theory to classify
virtually all the other algebras over the ground ring.

Azumaya's algebras over a commutative ring  F  were required to
satisfy two conditions:

(1)   The same map  $A \otimes_F A^0 \to \mathrm{Hom}_F(A,A)$  mentioned in Section 2 is
available even if  F  is not a field.  Azumaya demanded that
this be an isomorphism.

(2)   Furthermore, he asked that  A  be a finitely generated free
F-module, so  $\mathrm{Hom}_F(A,A)$  is a matrix algebra over  F .

Such algebras are now called Azumaya algebras, except that the
modern definition replaces (2) by, "A  is a finitely generated, faithful, projective F-module." — a concept that was not yet known in 1951.
Azumaya proved most of the fundamental properties of these Azumaya
algebras, but it is no surprise that his best theorem concerns algebras
over a local ring, where his limitation to free modules instead of

projective modules is no limitation at all.

In 1960, Auslander and Goldman [8] gave the correct definition of an Azumaya algebra replacing Azumaya's freeness hypothesis by projectivity. And, along the same lines, they corrected his definition of the Brauer group of a commutative ring. Azumaya had used the same definition of similarity between Azumaya algebras as the classical one; translating matrix rings into endomorphism rings and "matrices with entries in $A$ " to the tensor product of $A$ and a matrix ring over $F$ , the definition in Section 2 becomes: $A$ is <u>similar</u> to $B$ if $A \otimes_F \text{Hom}_F(D,D)$ is isomorphic to $B \otimes_F \text{Hom}_F(E,E)$ for some finitely generated free modules $D$ and $E$ . For the correct definition, once again simply change "free" to "faithful and projective." Just as in Section 2, the similarity classes of Azumaya algebras form a group under tensor product, and this is the Brauer group of $F$ , denoted $\text{Br}(F)$ .

The Brauer group is a functor of $F$ . If you have a homomorphism of rings, $F \to K$ , it induces a group-homomorphism $\text{Br}(F) \to \text{Br}(K)$ by associating to each Azumaya F-algebra $A$ the K-algebra $A \otimes_F K$ .

Azumaya's major result is that this homomorphism is an isomorphism when $F$ is a Hensel local ring (or, less generally, a complete local ring) and $K$ is its residue field ( $K = F/I$ with $I$ the maximal ideal of $F$ ). Auslander and Goldman have an example to show this is false for general local rings, but there are other cases where $\text{Br}(F) \to \text{Br}(F/I)$ is an isomorphism, for example, when $F$ is any ring and $I$ any nilpotent ideal; or when $(F,I)$ is a "Hensel pair" [35], [23], [20].

A second natural ring homomorphism on which to try the functor
Br is the inclusion  $F \to F[X]$  of  $F$  into its polynomial ring  $F[X]$ .
Now  $Br(F) \to Br(F[X])$  is always a monomorphism for functorial reasons
alone, because if we map  $F[X]$  to  $F$  by setting  $X = 0$  in every
polynomial, the composite map  $F \to F[X] \to F$  is the identity.  Hence,
the composite map  $Br(F) \to Br(F[X]) \to Br(F)$  is the identity, which
proves  $Br(F) \to Br(F[X])$  is a monomorphism (and, incidentally,
$\alpha : Br(F[X]) \to Br(F)$  is an epimorphism and  $Br(F[X])$  is the direct
product of  $Br(F)$  and  $\ker \alpha$ ).  Auslander and Goldman proved that,
when  $F$  is a field,  $Br(F) \to Br(F[X])$  is an isomorphism if and only
if  $F$  is perfect ([8], 7.5).  Yuan [40] has put the Auslander-Goldman
theorem on Brauer groups of polynomial rings to good use to explain a
classical theorem of Witt on Brauer groups of fields.  For every field  $F$
complete in a discrete valuation, Witt has a homomorphism from  $Br(F)$
to a certain explicit group associated with the residue class field  $k$
of  $F$ , namely, the character group of the automorphism group of  $k$ .
Witt has a formula for the kernel of this homomorphism if  $k$  is
perfect; it is  $Br(k)$ .  Yuan showed that, even if  $k$  is not perfect,
provided  $F$  and  $k$  have the same characteristic, the correct formula
for the kernel is  $Br(k[X])$ .  The Auslander-Goldman theorem says that
Yuan reduces to Witt when  $k$  is perfect.  As far as I know, the correct
formula remains to be found in the case of general  $k$  when the
characteristics are not equal.

Even if  $F$  is a perfect field,  $Br(F) \to Br(F[X_1,\ldots,X_n])$  need
not be an isomorphism when  $n > 1$ , ([28], 3.11).  However, there are

several theorems that assert this isomorphism for special rings  F ,
for example, regular domains of characteristic zero ([8], 7.7), or
affine algebras of (geometric) dimension one over a field of characteristic zero, or finite Z-algebras which are free as Z-modules, ([28], 3.8
and 3.9).

A third ring homomorphism that must come to mind is the inclusion
of an integral domain  F  into its field of fractions,  K .  Auslander
and Goldman ([8], 7.2) showed that the resulting map  Br(F) → Br(K)  is
a monomorphism if  F  is a regular domain (i.e., if all its localizations
at maximal ideals are regular local rings); this regularity hypothesis
cannot be dropped ([8], 8.5).  The map is rarely an isomorphism.  For
example, the Brauer group of the ring of ordinary integers is zero, but,
of course, the Brauer group of the field of rational numbers is not.

Much more is known about this homomorphism now.  D. S. Rim [33]
and B. Auslander ([6], Ch.2) found formulas for the kernel of the homomorphism when  F  is a domain that is not regular but is Noetherian
and integrally closed.  Childs, Garfinkel and Orzech have used similar
formulas to produce integrally closed Noetherian rings for which the
map of Brauer groups is not a monomorphism [19] and Childs has used them
to produce examples of other Noetherian domains for which the map of
Brauer groups is a monomorphism [18].

Parallel to this question, "Is  Br(F) → Br(K)  a monomorphism when
K  is the field of fractions of the domain  F ?" are two other questions.
Suppose  Br(F) → Br(K)  is a monomorphism; what is its image?  If  F  is
a Dedekind ring (which is regular) or, more generally a regular ring of

Krull dimension at most two, one answer is $\cap_p Br(F_p)$ , the intersection

taken over all minimal nonzero prime ideals $p$ of $F$ (i.e., all

nonzero primes in the Dedekind case; $F_p$ is the local ring of $F$ at

$p$ ). The intersection is to be understood as the intersection of

subgroups of $Br(K)$ , since each $F_p$ is also regular with field of

fractions $K$ , and so all the Brauer groups in question are embedded

in $Br(K)$ . ([8], 7.4.)

   A second question: Even if $Br(F) \to Br(K)$ is not a monomorphism,

perhaps $Br(F) \to Br(F_p)$ may be, that is, every Azuyama F-algebra

split by $F_p$ is already split by $F$ ; or, even more likely, every

algebra split by every $F_p$ must be already split by $F$ . This last

asks if $Br(F) \to \Pi_p Br(F_p)$ is a monomorphism. This is discussed in

([24], II) and [7]. For affine rings of dimension one it is true,

but there are counter-examples for integrally closed affine rings of

dimension two [16].

   With this much activity on the embedding of the Brauer of group of

a domain, it is perhaps not so surprising that quite a few Brauer groups

of domains are known. For example, if $F$ is the ring of algebraic

integers in a (finite) algebraic number field, then $Br(F)$ is a

direct product of cyclic groups of order $2$ ; the number of direct

factors is one less than the number of monomorphisms of $F$ into the

field of real numbers, ([24], III, 2.4). (This implies $Br(Z) = 0$ .)

If $F$ is the ring of a curve over an algebraically closed field $k$ ,

that is, if $F = k[X,Y]/I$ for some nonzero prime $I$ , then $Br(F) = 0$

([24], III), (Ch.2,p.86). (For nonsingular curves, F is regular, so
Br(F) = 0 can be deduced from Auslander-Goldman's Br(F) ⊂ Br(K) and
Tsen's Br(K) = 0 .) Then DeMeyer has used this, together with some
interesting Mayer-Vietoris sequences to compute the Brauer groups of the
rings of many other curves. For example, the Brauer group of the ring
of a real curve is a direct product of cyclic groups of order 2 [21].

Nevertheless, I think it is fair to say that, in general, Brauer
groups of rings are not well understood. There are virtually no general
facts, except that the Brauer group is always a torsion group ([29],
IV.6.1). If F is a ring of characteristic p , then Br(F) is also
p-divisible [36], [30]. In fact, the theory of p-algebras over fields
has been successfully extended to algebras over rings of prime
characteristic. [36], [41].

One of the techniques that has been quite successful in computing
some of these Brauer groups of rings has been the use of a Mayer-Vietoris
sequence. We refer to [15] and [28] for good expositions of this device.
Other techniques are more explicitly homological, directly generalizing
those of Section 3. We pass to these techniques in the next section.

§5. **More Cohomology**. The homology in Section 3 is as available when
F is a ring as when F is a field. If G is a group of automorphisms
of K over F , then $H^n(G,U(K))$ is still defined since the cohomology
in question is just group cohomology which needs nothing more than that
the group U(K) of units of K is a G-module, and it is. Amitsur

cohomology is also defined for arbitrary ring extensions  K  over  F .
And, if  K  is a Galois extension of  F  in the sense defined by
Auslander and Goldman ([8], Appendix) and exploited by Chase, Harrison
and Rosenberg [13] then the "Galois cohomology"  $H^n(G,U(K))$  is
isomorphic to the Amitsur  $H^n(K/F,U)$ , exactly as before.  Furthermore,
under special circumstances,  $H^2(K/F,U)$  is isomorphic to  $Br(K/F)$ =
$ker[Br(F) \to Br(K)]$ .  The first such theorem after Amitsur is [34],
followed by others, mainly deducible from [14] or from similar theorems,
one of which we shall describe momentarily.

Unfortunately, these special circumstances are too special.  Except
when  F  is a field or a local ring or other quite restricted kind of
ring,  $Br(F)$  is the union of  $Br(K/F)$'s  only if you allow  K  to range
over extensions of  F  which are not finite, and over finite extensions
for which  $Br(K/F)$  is not isomorphic to  $H^2(K/F,U)$ .

This failure of isomorphisms has been repaired by several mathematicians.  Chase and Rosenberg [14] found a seven-term exact sequence
of groups, one arrow of which is a homomorphism  $H^2(K/F,U) \to Br(K/F)$ ,
under hypotheses still including the unwanted finiteness of  K .  However, this exact sequence does provide a map and allows the possibility
of computing its kernel and cokernel.  Other mathematicians produced
similar maps and embedded them in short, medium or long exact sequences,
[17], [25], [27].

All these exact sequences involve another abelian group  $Pic(F)$ ,
the **Picard group** of  F , which is defined for every commutative ring
F , and which holds the clue to whether  $Br(K/F)$  will be isomorphic to

$H^2(K/F,U)$ . There are several equivalent definitions of $Pic(F)$ ; the
one closest to the spirit of the Azumaya definition of Azumaya algebra
is this:  each element of $Pic(F)$ is an isomorphism class of invertible
F-modules; a module  P  is **invertible** if  $P \otimes_F Q \cong F$  for some module
Q  ([10], Ch.II,§5,No.4).  The group operation in  $Pic(F)$ is $\otimes_F$ .  Equivalently, the invertible modules are those finitely generated F-modules
which localize to free, rank one modules at every maximal ideal of  F
(the rank one projective modules).  This group  $Pic(F)$ was well-known
in number theory already in the nineteenth century.  For the ring of
integers in an algebraic number field (or for any Dedekind ring) a
module is invertible if and only if it is isomorphic to an ideal; and
two ideals are isomorphic if and only if one is a principal ideal times
the other.  Thus the Picard group is the ideal class group of algebraic
number theory.  It vanishes exactly when the Dedekind ring in question
is a principal ideal domain.

Then  Pic , like  U  or  Br , is a functor from commutative rings
to abelian groups.  We can simply substitute  Pic  for  U  everywhere in
the definition of Amitsur cohomology, to get new Amitsur cohomology
groups,  $\ker[Pic(K^{n+1}) \to Pic(K^{n+2})]/Im[Pic(K^n) \to Pic(K^{n+1})]$ , called
$H^n(K/F,Pic)$ .  In these terms, we can give a long exact sequence which
may be considered the best repair to date of the isomorphism between
the Brauer group and the second cohomology [39]:

(\*)                    $0 \to H^1(K/F,U) \to E_1 \to H^0(K/F,Pic) \to \cdots$

                         $\to H^n(K/F,U) \to E_n \to H^{n-1}(K/F,Pic) \to \cdots$

where the E's are groups which we shall describe explicitly in a

moment.  Along with this exact sequence come two homomorphisms

$Pic(F) \to E_1$  and  $Br(K/F) \to E_2$ .  All this is defined for arbitrary

commutative ring extensions  $F \to K$ .  Moreover, if  $K$  is faithfully

flat over  $F$  then  $Pic(F) \to E_1$  is an isomorphism and  $Br(K/F) \to E_2$

is a monomorphism; and the latter is an isomorphism if, in addition,

there exists a K-module which is finitely generated, faithful and

projective over  $F$  (Chase and Rosenberg assume that  $K$  is such a

module).  In the case of a Galois (hence, finite) extension  $K$  of  $F$ ,

Childs [17] has this same sequence and Hattori has a closely similar,

but different one [25].

The definition of  $E_n$  is a K-theoretic refinement of  $H^{n-1}(K/F, Pic)$ .

Instead of an element (isomorphism class) in  $Pic(K^n)$  we take a single,

invertible $K^n$-module  $P$ .  Define its "faces"  $\varepsilon_i P$  ($i = 0, \ldots, n$)  as

$P \otimes_{K^n} K^{n+1}$  where  $K^{n+1}$  is a $K^n$-module by virtue of the ring homomorphism

$\varepsilon_i : K^n \to K^{n+1}$  described in Section 3 (insert  $\otimes 1 \otimes$   as the $(i + 1)$-st

factor in the tensor product).  Define  $(\varepsilon_i P)^{-1}$  as the inverse of the

invertible module, or, more precisely, its dual,  $Hom_{K^{n+1}}(\varepsilon_i P, K^{n+1})$ .

Define the boundary  $\delta P$  as  $\varepsilon_0 P \otimes (\varepsilon_1 P)^{-1} \otimes \cdots \otimes (\varepsilon_n P)^{\pm 1}$ .  Consider

pairs  $(P, \beta)$  with  $\beta$  an isomorphism of  $K^{n+1}$-modules  $\beta : \delta P \to K^{n+1}$ .

The isomorphism classes of such pairs form a group of which  $E_n$  is

a subquotient:  We use only those pairs for which  $\delta\beta : \delta\delta P \to \delta K^{n+1} = K^{n+2}$

is the "canonical" isomorphism  $\lambda_P : \delta\delta P \to K^{n+2}$  (this is like the

topologists'  $\delta\delta = 0$ ), and reduce this subgroup of pairs modulo the pairs $(\delta Q, \lambda_Q)$ where $Q$ is an invertible $K^{n-1}$-module.

In general, then, we do not even have a map between $Br(K/F)$ and $H^2(K/F,U)$ . Instead, both groups map to the same $E_2$ , and, under the very best circumstances, the two maps are isomorphisms. In a sense, this relieves the tension between two approaches. One, as in Section 3, maps $H^2$ to $Br$ by associating to each cocycle a crossed-product or analogous algebra, which defines a class in $Br(K/F)$ . This approach even suggests reversing one of the developments in Section 3, going back to the algebras with $K$ as maximal commutative subalgebra, like the crossed-products, instead of the algebras split by $K$ , [14], [27], [25].

The other approach maps $Br(K/F)$ to some kind of second cohomology (like $E_2$ in (*), or like $H^2(K/F,U)$ if the hypotheses are strong enough to make this isomorphic to $E_2$ ). This approach seems necessary if one is to consider $K$'s which are not finite; and, of course, it comes up even with the opposite approach when one proves that a map of cohomology into $Br(K/F)$ is surjective. It is essentially descent theory. To sketch the definition of this map, use the notation $\varepsilon_i$ , as in the definition of $E_n$ above, to denote the ring homomorphism $K^n \to K^{n+1}$ as well as the corresponding functors from $K^n$-modules (resp., $K^n$-algebras) to $K^{n+1}$-modules (resp., $K^{n+1}$-algebras). Then, if $A$ is an Azumaya F-algebra split by $K$ , pick a splitting isomorphism $\alpha$ from a split K-algebra $E$ ( $= End_K(V)$ with $V$ finitely generated, faithful and projective over $K$ ) to $A \otimes_F K = \varepsilon_0 A$ . We get two $K^2$-algebra isomorphisms $\varepsilon_i\alpha : \varepsilon_i E \to \varepsilon_i \varepsilon_0 A$ (i = 0,1) ; but the usual face identities say $\varepsilon_0\varepsilon_0 = \varepsilon_1\varepsilon_0$ (in this sketch we hide a lot of coherence and natural isomorphism behind equal signs), so $\varepsilon_1 \alpha^{-1} \circ \varepsilon_0 \alpha$ is a $K^2$-algebra

isomorphism from $\varepsilon_0 E$ to $\varepsilon_1 E$. Such an isomorphism of split algebras

gives an invertible $K^2$-module P, thus: the isomorphism makes every

$\varepsilon_1 E$-module (in particular, $\varepsilon_1 V$) an $\varepsilon_0 E$-module. But $\varepsilon_0 E = \text{End}_{K^2}(\varepsilon_0 V)$

and it is known that all modules over such a ring are of the form

$\varepsilon_0 V \otimes_{K^2} P$ for some invertible module P; this is just the Morita

theorem.

So P "is" the boundary of $\alpha$. It is not surprising that $\delta P$

is isomorphic to the unit invertible module over $K^3$, namely, $K^3$

itself. Pick such an isomorphism $\beta$ and get a pair $(P, \beta)$ determining

an element of $E_2$. After all equivalences are counted up, the several

choices involved in this construction lead to a single-valued map

$Br(K/F) \to E_2$.

Grothendieck noticed that one can set up a dictionary which trans-

lates all the steps in computing Amitsur cohomology into corresponding

steps in computing Čech cohomology. (And Grothendieck then invented the

Grothendieck topology or site which includes both as special cases [5].)

Amitsur uses a ring extension $F \to K$; Čech needs a covering $\{U_\alpha\}$ of

a topological space X. Grothendieck calls K a covering of F, and

he thinks of a covering of a topological space X as a map of the dis-

joint union of the $U_\alpha$ to X, where each $U_\alpha$ maps into X by the

inclusion map. Of course, the dictionary reverses all arrows since

Amitsur is doing algebra and Čech is doing geometry.

Next, Amitsur constructs $K^2$, $K^3$, etc. while Čech builds the

coverings $\{U_\alpha \cap U_\beta\}$, $\{U_\alpha \cap U_\beta \cap U_\gamma\}$, etc. but in category terms $K^2$

is the coproduct of extensions (it satisfies the universal property:
there are two F-algebra homomorphisms of $K$ into $K^2$ and any other
pair of homomorphisms $K \to L$ induces a unique map $K^2 \to L$ ) while
the intersection $U_\alpha \cap U_\beta$ of subsets of $X$ is the category product
(satisfies the universal property: $U_\alpha \cap U_\beta$ is included in $U_\alpha$ and in
$U_\beta$ , and any other set which is included in $U_\alpha$ and $U_\beta$ is included
in the intersection).

Both Amitsur and Čech have face maps which correspond exactly, and,
after applying a functor from coverings to abelian groups (e.g., U or P.
Čech cohomology can use any "presheaf of coefficients") these face maps
can be combined to give boundary maps with resulting cohomology groups.

Thus Amitsur cohomology is the strict analog of the Čech cohomology
of a covering. In fact, if $K$ is a "Zariski cover" of $F$ , this is more
than an analogy. A **Zariski** **cover** of $F$ is the direct product
$K = \Pi F_f$ of rings of fractions of $F$ (allowable denominators in $F_f$
are the powers of the element $f$ ) where the product is taken over a
finite set of elements $\{f\}$ of $F$ which generate the unit ideal. In
geometric terms, the set $\{f\}$ corresponds exactly to an open covering
of $\mathrm{Spec}(F)$ , and the resulting $H^n(K/F,U)$ is exactly the Čech cohomology
of this covering of the topological space $\mathrm{Spec}(F)$ , with coefficients
in the sheaf $U$ .

Then the topologist recognizes that to get properties of the space
instead of properties of the covering, you need to take the direct limit
of the cohomology of the covering as the covering gets arbitrarily fine.
The analog for the algebraist is to take the direct limit of $H^n(K/F,U)$

as  K  "goes to infinity," to get information about  F  itself.  This
works perfectly well:  When  $K \to L$  is a homomorphism of F-algebras, we
get a group homomorphism  $H^n(K/F,U) \to H^n(L/F,U)$  and the direct limit of
these groups we call  $H^n(F,U)$ .

    If  F  is a field, then  $H^2(K/F,U)$  is  $Br(K/F)$ , a subgroup of
$Br(F)$ ; and the direct limit is just the union of subgroups of  $Br(F)$ .
We saw in Section 3 that every Azumaya algebra over  F  has a splitting
field  K , which asserts that this direct limit is exactly  $Br(F)$ .

    This leads to the expectation that, even though  $Br(K/F)$  and
$H^2(K/F,U)$  are not the same, in the limit  $H^2(F,U)$  might equal  $Br(F)$ .
Following Grothendieck, we must be more precise about which K's are used
in taking the direct limit; there are several classes of extensions that
are reasonable candidates.  Everyone agrees that the K's should all be
faithfully flat, (because of the repeated need for descent theory), but
the class of all faithfully flat extensions may be too big; on the other
hand, you need a class which is large enough that every Azumaya algebra is
split by some extension in the class.  The best compromise may be the class
of "étale" extensions which includes all the Zariski covers and all
"strongly separable" extensions of Zariski covers, but not much more.

    However, even using this best class of K's, the limit  $H^2(F,U)$  is
not equal to the Brauer group of  F .  There is a monomorphism of the latter
to the former.  It is not always an isomorphism because  $Br(F)$  is always
a torsion group, but there are examples where  $H^2(F,U)$  is not.  Of course,
if  F  is a field,  $Br(F)$  is isomorphic to this limit of cohomology
groups, and the same is true for more general F's, for example, regular

domains of Krull dimension  1 .  A full analysis of the difference

between these two groups is still not available.  ([24], II.)  For

example, is the Brauer group of  F  always the full torsion subgroup of

$H^2(F,U)$ ?

The long exact sequence (*) is not much extra help, here.  If we

take the direct limit of this sequence over a suitable class of K's

(for example, the étale ones) then the limit of  $H^n(K/F,Pic)$  is zero

and the sequence degenerates into a collection of isomorphisms

$\lim_K(E_n) \cong H^n(F,U)$ , which at best gives an alternative version of

$H^n(F,U)$ .

The whole cohomological approach leaves one with the automatic

question:  If  $H^0(K/F,U)$  is  U  and  $H^1$  is usually  Pic  and  $H^2$  or

a suitable substitute is  Br , what is  $H^3(K/F,U)$ , or  $H^3(F,U)$ , or

$E_3$ ?  Childs has some clues [17] and there is a Teichmüller cocycle

which is another, but the question is very far from solved.

We conclude with a nonhomological description of  U , Pic  and

Br , which also suggests that there should be a next term in this series.

For any commutative ring  F , we defined  U(F)  to be the group of

**invertible elements** of  F , and  Pic(F)  to be the group of **isomorphism**

**classes** of **invertible modules** over  F .  Moreover, the identity element

of  Pic(F)  is the class of  F , and the automorphism group of this

module is  U(F) .  As a direct analogy,  Br(F)  can be defined as the

group of **Morita equivalence classes** of **invertible F-algebras**.  We explain

this momentarily.  However, the identity element of this group is the

class of  F , and the automorphism group of  F , in the category involved in the following explanation, is exactly $\text{Pic}(F)$ .

A Morita equivalence from an F-algebra $A$ to an F-algebra $B$ is a category equivalence over $F$ from the category of A-modules to the category of B-modules. The Morita theorem says that such an equivalence is simply $(\cdot) \otimes_A P$ for some $(A \otimes_F B^0)$-module, $P$ which is invertible in the sense that there exists a $(B \otimes_F A^0)$-module $Q$ with $P \otimes_B Q \cong A$ and $Q \otimes_A P \cong B$ . Thus, one may think of a Morita equivalence as "being" an invertible bimodule, $P$ . If $A = B = F$ , the Morita equivalences "are" just the invertible modules that make up $\text{Pic}(F)$ .

We consider the category of F-algebras where the "maps" in the category are the Morita equivalences. Of course, every algebra isomorphism "is" such a map. This category has a product, namely, tensor product over $F$ . And the algebra $F$ is the identity element for this product. An algebra $A$ is "isomorphic to $F$ " in this category if there exists a Morita equivalence from $F$ to $A$ . We say " $A$ is Morita equivalent to $F$ ." By the Morita theorem, these are exactly the split Azumaya algebras, $\text{End}_F(V)$ . We say that an algebra $A$ is invertible if there exists an algebra $B$ such that $A \otimes_F B$ is Morita equivalent to $F$ . It is easy to show that an algebra is invertible if and only if it is an Azumaya algebra. Further, two Azumaya algebras are Morita equivalent if and only if they are similar, or Brauer equivalent as in the definition in Section 4. So $\text{Br}(F)$ is indeed the group of Morita equivalence classes of invertible F-algebras.

Then what is the next category?

## REFERENCES

1.  Albert, A. A.  "A construction of noncyclic normal division algebra
    Bull. A.M.S. 38(1932), 449-456.

2.  _____.  Structure of Algebras, A.M.S. Colloquium Pub. Vol. 2
    Providence, R.I., 1961.

3.  Amitsur, S. A.  "Simple algebras and cohomology groups of arbitrary
    fields," Trans. A.M.S. 90(1959), 73-112.

4.  _____.  "On central division algebras," Israel Math. J. 12(1
    408-420.

5.  Artin, M.  Grothendieck Topologies, Harvard Lecture Notes, 1962.

6.  Auslander, B.  "The Brauer group of a ringed space," J. Alg. 4(1966)
    220-273.

7.  _____.  "Central separable algebras which are locally endo-
    morphism rings of free modules," Proc. A.M.S. 30(1971), 395-404.

8.  Auslander, M. and O. Goldman.  "The Brauer group of a commutative
    ring," Trans. A.M.S. 97(1960), 367-409.

9.  Azumaya, G.  "On maximally central algebras," Nagoya Math. J. 2(1951
    119-150.

10. Bourbaki, N.  Algèbre Commutative, Actualités Scientifiques, No. 1290

11. Brauer, R.  "Über die algebraische Structur von Schiefkörpern," J.
    f. d. reine u. angewandte Math.166(1932), 241-252.

12. Brauer, R., H. Hasse and E. Noether.  "Beweis eines Hauptsatz in der
    Theorie der Algebren," J. f. d. reine u. angewandte Math. 167(1932),
    399-404.

13. Chase, S. U., D. K. Harrison and A. Rosenberg. "Galois theory and Galois cohomology of commutative rings," Memoirs A.M.S. 52(1968), 15-33.

14. Chase, S. U. and A. Rosenberg. "Amitsur cohomology and the Brauer group," Memoirs A.M.S. 52(1968), 34-79.

15. Childs, L. N. "Mayer-Vietoris sequences and Brauer groups of non-normal domains," Trans. A.M.S. 196(1974), 51-67.

16. _____. "Brauer groups of affine rings," Proc. Okla. Conf. (ed. by B. R. McDonald, A. R. Magid, K. C. Smith), Marcel Dekker Lecture Notes No. 7, New York, 1974.

17. _____. "On normal Azumaya algebras and the Teichmüller cocycle map," J. Alg. 23(1972), 1-17.

18. _____. "On Brauer groups of some normal local rings," Northwestern Univ. Conf. on Brauer Groups, Oct. 1975, Springer-Verlag Lecture Notes, to appear.

19. Childs, L. N., G. Garfinkel and M. Orzech. "On the Brauer group and factoriality of normal domains," Queen's Mathematical Preprints, 1974-9.

20. DeMeyer, F. "The Brauer group of a ring modulo an ideal," Rocky Mt. J. Math., to appear.

21. _____. "The Brauer group of a real curve," to appear.

22. DeMeyer, F. and E. Ingraham. Separable Algebras over Commutative Rings, Springer Lecture Notes, No. 181, New York, 1971.

23. Greco, S. "Algebras over nonlocal Hensel rings II," J. Alg. 13(1969), 48-56.

24. Grothendieck, A. Le Groupe de Brauer, I, II, III, in Dix exposés sur la cohomologie des schémas, North-Holland Pub. Co., Amsterdam, 1

25. Hattori, A. "Certain cohomology associated with Galois extensions of commutative rings," Sci. Papers, Coll. Gen. Ed., Univ. Tokyo 24(1974), 79-91.

26. Herstein, I. N. Noncommutative Rings, Carus Monograph No. 15, M.A.A., Wiley, New York, 1968.

27. Kanzaki, T. "On generalized crossed product and Brauer group," Osaka J. Math. 5(1968), 175-188.

28. Knus, M.-A. and M. Ojanguren. "A Mayer-Vietoris sequence for the Brauer group," J. Pure and Applied Algebra 5(1974), 345-360.

29. _____. Théorie de la descente et Algèbres d'Azumaya, Springer Lecture Notes No. 389, New York, 1974.

30. Knus, M.-A., M. Ojanguren and D. J. Saltman. "On Brauer groups in characteristic p ," Northwestern Univ. Conf. on Brauer Groups, Oct., 1975, Springer-Verlag Lecture Notes, to appear.

31. Lang, S. "On quasi algebraic closure," Ann. of Math. 55(1952), 373-390.

32. Orzech, M. and C. Small. Brauer Groups of Commutative Rings, Marcel Dekker Lecture Notes No. 11, Marcel Dekker, New York, 1975.

33. Rim, D. S. "An exact sequence in Galois cohomology," Proc. A.M.S. 16(1965), 837-840.

34. Rosenberg, A. and D. Zelinsky. "On Amitsur's complex," Trans. A.M.S. 97(1960), 327-356.

35.  Roy, A. and R. Sridharan.  "Derivations in Azumaya algebras," J.
     Math. Kyoto U. 7(1967), 161-167.

36.  Saltman, D. J.  "Azumaya algebras over rings of characteristic  p ,"
     Yale Univ. Dissertation, 1976.

37.  Serre, J.-P.  Corps Locaux, Hermann et Cie, Paris, 1968.

38.  Tsen, C. C.  "Divisionsalgebren über Funktionenkörpern,"
     Göttingen Nachrichten II, No. 48(1933), 335-339.

39.  Villamayor, O. E. and D. Zelinsky.  "Brauer groups and Amitsur
     cohomology for arbitrary commutative ring extensions," J. Pure
     and Applied Algebra, to appear.

40.  Yuan, S.  "On the Brauer groups of local fields," Ann. of Math.
     82(1965), 434-444.

41.  _____.  "Central separable algebras with purely inseparable
     splitting fields of exponent one," Trans. A.M.S. 153(1971),
     427-450.

# THE STRUCTURE OF THE UNIT GROUP OF GROUP RINGS

R. Keith Dennis[1]

Cornell University

§1.  **Introduction**.  For  R  an arbitrary associative ring with unit and
G  an arbitrary group,  RG  will denote the group ring and  U(RG)  its
group of units.  In the following we ask a number of natural questions
about the structure of  U(RG)  and give references to the literature for
further information.  We have been quite brief in the first three sections as there is an extensive discussion of these topics already in
print.  Even though the Jacobson radical of a ring is quite relevant to
the study of the group of units of the ring, we have omitted that
particular topic from our discussion.  One should consult [62], [63],
and [94] for more information concerning the Jacobson radical of group
rings.

Information on  U(RG)  can be found in the two excellent survey
articles on group rings [63, §10, pp.94-97], [94, §2, pp.7-18] and two
books [62, §26, pp.110-114], [17].  According to Passman ("What is a
group ring?", Amer. Math. Monthly 83(1976), 173-184) a second volume of

---

[1]Partially supported by NSF-MPS 73-04876.

Bovdi's book [17] is anticipated and an expanded and updated version of [62] entitled The Algebraic Structure of Group Rings is in preparation.

The structure of U(RG) should be considered as a special case of the structure of the unit group of an arbitrary ring. See [13], [28], [47], [51], and [93]

A number of problems dealing with the structure of U(RG) arose from algebraic K-theory and can, in fact, be studied by its techniques. One should see [3], [4, Chapter XI, §7; Chapter XII, §10], [30], [33], [52], [58], [60], [86], [87], [88], and [91]. These particular questions in algebraic K-theory had their origin in topology via the Whitehead group of simple homotopy theory. A few references to the topological content of our questions can be found in [19], [20], [52], [59], and [60]

We denote by $\varepsilon : RG \to R$ the ring homomorphism which is the identity on R and sends group elements to 1. V(RG) denotes the kernel of $\varepsilon$ restricted to U(RG). This yields the decomposition

$$U(RG) = V(RG) \rtimes U(R)$$

(a semi-direct product which becomes a direct product in case R is commutative).

I would like to thank H. Bass, E. Formanek, D. S. Passman, S. K. Sehgal, and W. van der Kallen for making a number of helpful remarks on a preliminary version of this paper.

§2.  One-sided Units.

Problem 1.  If every one-sided unit of  R  is actually a unit, must
every one-sided unit of  RG  be a unit?  ([46, Problem 23] and [45,
p.122].)

Problem 2.  Find necessary and sufficient conditions on  R  in order
that every one-sided unit of  $M_n(RG)$  is a unit for all  $n \geq 1$ .

    Note that Shepherdson [84] has given an example of a domain  R
with every one-sided unit actually a unit and having one-sided units
that are not units in all matrix rings  $M_n(R)$  for  $n \geq 2$ .
    If  R  is a commutative domain of characteristic  0 , every one-
sided unit of  $M_n(RG)$  is actually a unit [45, p.122], [53] (see also
[34, pp.33-36]).
    Also consult [42], [49], [66] and [80].

§3.  Group Properties.  Let  P  denote a property of groups.

Problem 3.  Determine necessary and sufficient conditions on  R  and  G
in order that  U(RG)  or  V(RG)  have property  P .  In particular, one
would like the answer when  P  denotes solvable, nilpotent or metabelian.
    For information on the solvable case see [6], [7], [13], [31], [54],
[56], [80], [81] and [94, p.18].  For information on the nilpotent case

see [8], [27], [36], [37], [38], [55], [67], [80], [83] and [86, pp. 17-18].

§4.  Isomorphisms.

Problem 4.  Given a ring  R  and groups  G  and  H  determine when
U(RG) $\approx$ U(RH) .

Note that there exist two distinct finite groups with isomorphic group rings over any field [25].  Other examples are given in [61].  In [21],  U(RG)  is used as a tool to study the isomorphism problem for group rings.

§5.  Splittings.

Problem 5.  Find necessary and sufficient conditions on  R  and  G  so that there exists a homomorphism  s : U(RG) $\rightarrow$ G  left inverse to the inclusion  G $\rightarrow$ U(RG) .

This problem seems to have been considered only recently [26]. (According to the Zentralblatt review, [44] considers such questions.) The following theorem appears in [26].

<u>Theorem.</u>  Let  G  be an arbitrary group.  Then

(*)                           $s(N) = \left[ \pi \bar{g}^{-n(g)} \right]^{\varepsilon(N^{-1})}$

where  N  is the class of  $\Sigma\, n(g)g$ , defines a homomorphism

$s : U(ZG)^{ab} \to G^{ab}$  left inverse to the homomorphism  $G^{ab} \to U(ZG)^{ab}$ .

Let  t  be  s  restricted to  V(ZG) .  Then  $U(ZG)^{ab} = U(Z) \times G^{ab} \times \ker t$ .

In particular, if  G  is abelian, then  $U(ZG) = U(Z) \times G \times \ker t$ .


The above results can be generalized provided some sense can be
made of formula (*).  For example, if  G  has the structure of an R-
module so that  rs - sr  acts as  0  on  G  for all  r  and  s  in  R .
[One can weaken the hypothesis "module" to "semi-module."]  We reformulate
Problem 5 in the following manner.


<u>Problem 6.</u>  Given a ring  R  determine the class  $\underline{G}(R)$  (respectively,
$\underline{G}(R)$ ) of all groups (respectively, of all abelian groups) so that the
inclusion  $G \to U(RG)$  splits.


In [26] it is shown that  $\underline{G}(R)$  is
(1)   the class of all abelian groups if  R  is a subring of the complex
      numbers which are integral over  Z ,
(2)   the class of all groups of the form  $D \oplus H$  where  D  is divisible
      and  H  is torsion-free if  R  is a field containing  $\tilde{Q}$  ( Q  with
      all roots of unity adjoined),
(3)   the class of all groups of the form  $D \oplus H \oplus E$  where  D  is

divisible, H is torsion-free and E is an elementary 2-group

if R is a real-closed field (e.g., $\mathbb{R}$ ),

(4) is contained in the class of all groups of the form D $\oplus$ H where

D is divisible and the only torsion in H is p-torsion in case

R is a field containing $\bar{\mathbb{F}}_p$ (the algebraic closure of the field

with p elements).

For even quite reasonable rings R , these questions can be exceedin̄

difficult. For example, the cyclic group of prime order p is in $\underline{G}(\mathbb{F}_2)$

if and only if $p^2$ does not divide $2^{p-1} - 1$ . Now $p^2$ divides $2^{p-1} -$

for p = 1093 , p = 3511 and for no other primes $p < 3 \times 10^9$ . A

number of other questions are asked in [26] and it is wildly conjectured

that $\underline{G}(Q)$ is the class of all abelian groups and $\underline{G}(\bar{\mathbb{F}}_p)$ is the class

of all groups of the form D $\oplus$ H where D is divisible and the only

torsion in H is p-torsion.

The following lemma is quite useful to investigate the splitting

problem for non-abelian groups.

<u>Lemma</u>. G is in $\underline{G}(Z)$ if and only if there exists a ring S and a

split homomorphism G → U(S) .

For example, $S_3 = GL_2(\mathbb{F}_2)$ , the quaternion group of order 8

(the unit group of Z[i,j,k]) , and the dihedral group of order 8 (the

unit group of the ring of $3 \times 3$ upper triangular matrices over $\mathbb{F}_2$ )

are in $\underline{G}(Z)$ . One can also show that $A_4$ and the dihedral groups of orders  10  and  12  are in  $\underline{G}(Z)$ .

This lemma can also be used to give a simple proof of the theorem stated at the beginning of this section.

§6.  Normal Subgroups.

Problem 7.  Classify the normal subgroups of  $U(RG)$  and  $V(RG)$ .

Problem 8.  Which subgroups  H  of  G  are normal subgroups of  $V(RG)$ ?

Bovdi [15], [16] and Bovdi, Hripta [18] have some results on Problem 7 (see [91, Theorem 2.39, Theorem 2.40]).

Pearson [64], [65] has shown that if  char $R \neq 0$  and  H  is a subgroup of the finite group  G , then  H  is a normal subgroup of  $V(RG)$  if and only if either  H  is a subgroup of the center of  G  or  $R = \mathbb{F}_2$ ,  $G = S_3$  and  $H = A_3$  or  $S_3$ .

It easily follows from [29] that if  R  is a ring in which  p  is nilpotent, and  G  is a locally finite p-group, then  H  is normal in  $V(RG)$  if and only if  H  is central in  G .

If  G  is a finite p-group and  F  is a field of characteristic p , then the normalizer of  G  in  $U(FG)$  is  GZ  where  Z  is the center of  $U(FG)$  [22].

Berman [10] (see [64]) has shown that for  G  finite,  G  is normal in  $U(ZG)$  if and only if  G  is abelian or the direct product of the quaternion group and an elementary 2-group.

In [2, Lemma 5] one finds the following result.

**Lemma.** If ZG is a domain and u in U(ZG) is such that $uGu^{-1} = G$,
then u is in ±G .

§7. **Trivial Units.** A unit of RG is called **trivial** if it is of the
form ug , u in U(R) , g in G .

Problem 9. Determine when RG has only trivial units.

The following seem to be the major results:

**Theorem.** [63, Theorem 10.1]. Let F be a field and let G be a group
which is not torsion-free. Then FG has nontrivial units unless
$F = \mathbb{F}_2$ and $|G| = 2,3$ or $F = \mathbb{F}_3$ and $|G| = 2$ .

**Theorem.** (Higman). [35, Theorem 11, p.240]. (See [94, Theorem 2.38],
[10], [15], [40], [64] and [21, Theorem 2.3].) Let R = Z and let G
be a torsion group. Then every unit of ZG is trivial if and only if
G is either

(1) abelian of exponent dividing 4 ,

(2) abelian of exponent dividing 6 , or

(3) the direct product of the quaternion group and an elementary
2-group.

A group  G  is said to be a <u>unique</u> <u>product</u> <u>group</u> (u.p. group) if
given any two nonempty finite subsets  A  and  B  of  G , then there
exists at least one element  x  in  G  which has a unique representation
in the form  x = ab  with  a  in  A  and  b  in  B .

A group  G  is said to be a <u>two</u> <u>unique</u> <u>products</u> <u>group</u> (t.u.p. group)
if given any two nonempty finite subsets  A  and  B  of  G  with
$|A| + |B| > 2$ , then there exists at least two distinct elements  x  and
y  of  G  which have unique representations in the form  x = ab ,
y = cd  with  a  and  c  in  A  and  b  and  d  in  B.


<u>Theorem</u>.  [63, Lemma 10.2].  If  R  has no zero divisors and  G  is a
t.u.p. group, then  RG  has only trivial units.


<u>Theorem</u>.  (S. K. Sehgal).  [63, Theorem 10.5].  If  G  is a u.p. group
and  R  is a commutative domain of characteristic  0 , then  RG  has only
trivial units.


The following are examples of t.u.p. groups:
(1)   ordered and right-ordered groups [73], [62, Lemma 26.4] (free or
       locally nilpotent torsion-free groups are orderable;   ordered and
       right-ordered groups are not equivalent [92]);
(2)   locally indicable [35, Theorem 13, p.243] (A group is <u>locally</u>
       <u>indicable</u> if every finitely generated subgroup admits a surjective
       homomorphism to the infinite cyclic group.  If  H  is a normal
       subgroup of  G  and both  H  and  G/H  are locally indicable, then

so is  G .  The free product of two locally indicable groups is

locally indicable.  [35, Appendix]);

(3)  if  $1 = G_0 \lhd G_1 \lhd \cdots \lhd G_n = G$  is a subnormal series and

$G_i/G_{i-1}$  is torsion-free abelian [14] (If  G  is a finitely generated

torsion-free group with an abelian subgroup  A  of finite index

with  G/A  cyclic, then this applies [32].).

Note that  $Z \times_\phi Z$ ,  $\phi : Z \to \mathrm{Aut}(Z)$  the nontrivial homomorphism, is

an example of a t.u.p. group that is not orderable [62, p.113].

The following theorem tells how to obtain more examples of u.p.

and t.u.p. groups.

Theorem.  [62, Theorem 26.3].  If  G  is a group, then any of the fol-

lowing imply that  G  is a u.p. group (respectively, a t.u.p. group):

(1)  $H \lhd G$  and  H,G/H  are both u.p. groups (respectively, t.u.p.

groups),

(2)  $H_i \lhd G$  for all  i ,  $\cap H_i = 1$  and  $G/H_i$  is a u.p. group

(respectively, a t.u.p. group) for all  i ,

(3)  every nontrivial finitely generated subgroup of  G  can be

mapped onto a nontrivial u.p. group (respectively, a t.u.p.

group).

Remark.  It has been conjectured [94] that if  R  is a ring without

zero divisors and  G  is a torsion-free group then

(a)  RG  has no nontrivial idempotents,

(b)  RG  has ho zero divisors,

(c)   RG   has no nilpotent elements,

(d)   RG   has only trivial units.

It is shown in [94, Proposition 2.7] and in [63, p.95] that (d)

implies (c), (c) and (b) are equivalent, and (b) implies (a).   Thus

Problem 9 is the hardest of the lot.

§8.   Units of Finite Order.

Problem 10.   Determine all of the elements of finite order in  V(RG) .

Problem 11.   When is every element of finite order in  V(RG)   trivial?

Problem 12.   When is every element (subgroup) of finite order in  V(RG)

conjugate to an element (subgroup) of  G ?

Note that Problem 12 is closely related to Problem 8.

Problem 13.   Let  G  be a group of exponent  n .   When is every torsion

element of  V(RG)   annihilated by  n ?   (See [95].)

Note that if  G  is a finite group and  ZG  contains a nontrivial

unit of finite order, then it contains infinitely many nontrivial units

of finite order [50].   If  u  is a torsion unit of  ZG , G  finite, then

either  u = ±1  or the coefficient of  1  is  0  [10, Lemma 2], [21] or

[89]. The hypothesis that G is finite can be deleted [82, p.141].
It is conjectured (see [95]) that any unit of finite order in U(ZG) is
conjugate in U(QG) to an element of ±G .

The following theorems are some of the main results.


Theorem. (Higman). [35, Theorem 3, p.237]. If R is a ring of
algebraic integers and G is a (finite) abelian group, then every unit
of finite order in RG is trivial.
(The word "finite" can be deleted.)


Theorem. (Bovdi). [13]. If U(ZG) is torsion, then every unit is
trivial. (Hence G is determined by Higman's theorem. See Section 6
above.)


Theorem. (Hughes and Pearson). [39].

(1) $U(ZS_3)$ is isomorphic to the subgroup of $GL_2(Z)$ consisting
of all matrices such that the sum of the entries in the first
column is congruent modulo 3 to the sum of the entries of
the second column.

(2) The torsion elements of $U(ZS_3)$ have order 2 or 3 . Any
element of order 3 is conjugate to an element of $S_3$ . There
are two conjugacy classes of elements of order 2 represented
by (12) and $t = (12) + 3(13) - 3(23) - 3(123) + 3(132)$ .
These two elements are conjugate in $QS_3$ .

(3) Any maximal finite subgroup of $V(ZS_3)$ is conjugate to $S_3$

or  $\{1,t\}$ .

<u>Theorem</u>.  (Zassenhaus).  [95].  Let  $R$  be a commutative domain and let  $G$  be a finite group.  If no prime divisor of the order of  $G$  is a unit of  $R$ , then the order of any torsion element of  $V(RG)$  is a divisor of the exponent of  $G$ .

<u>Theorem</u>.  (Sehgal).  [82, p.142].  Let  $R$  be a ring of algebraic integers.  If  $p$  is a prime and there is an element of  $V(RG)$  of order  $p^n$ , then  $G$  contains an element of order  $p^n$ .  In particular, if  $G$  is torsion-free, so is  $V(RG)$ .

This theorem is also proved in [5].  For further results consult [40], [64] and [96].

§<u>9</u>.  <u>Integral Group Rings</u>.  Let  $G$  be an abelian group.  Then one has the canonical decomposition (see Section 5 above)

$$U(ZG) = U(Z) \times G \times F$$

where  $F$  is torsion-free.  (By Theorem 1 of [78] every unit of  $ZG$  is of the form  $gu$  where  $u$  is a unit of  $ZH$ ,  $H$  a finite subgroup of  $G$  and  $g$  is in  $G$ .  The only torsion in  $ZH$  is  $\pm H$  by Higman's theorem.)  If  $G$  is finite,  $F$  is free of rank  $(|G| + 1 + t_2 - 2\ell)/2$  where  $t_2$  is the number of elements of order  $2$  in  $G$  and  $\ell$  is the number of cyclic subgroups of $G$ [1, Theorem 4]. If $G$ is finitely generated, then $F$ is also free. Its rank can be computed as the number obtained from the preceding formula applied to the torsion subgroup of $G$ .

Problem 14. Find necessary and sufficient conditions on $G$ in order that $F$ be a free abelian group.

Problem 15. Explicitly give a collection of free generators for $F$ when $G$ is finite.

If $G$ is cyclic of order $5$ with generator $t$ , then $F$ has rank $1$ and Kaplansky has shown that $t + t^{-1} - 1$ (with inverse $t^2 + t^{-2} - 1$ is a free generator for $F$ [52, p.375].

For $G$ finite the rank of $F$ is computed by considering the map $U(ZG) \rightarrow U(QG)$ . Let $\zeta$ be a primitive p-th root of unity. Note that the map $U(ZG) \rightarrow U(Z[\zeta])$ given by sending the generator of $G$ to $\zeta$ is injective. For $p = 5$ the image of $\pm G$ and $t + t^{-1} - 1$ generates a subgroup of index $2$ . It is easy to see that the unit $\zeta + \zeta^{-1}$ is not in the image of $U(ZG)$ , hence Kaplansky's result. For $p = 7$ the image of $\pm G$ , $t + t^{-1} - 1$ and $t^2 + t^{-2} - 1$ generates a subgroup of index $3$ . As before the unit $\zeta + \zeta^{-1}$ is not in the image and we have found free generators for $F$ . On the basis of this scanty evidence one might be tempted to conjecture that for $G$ cyclic of order $p$ , $t^i + t^{-i} - 1$ , $i = 1,\ldots,(p - 3)/2$ , are free generators for $F$ and that the index of

the image of  U(ZG)  in  U(Z[ζ])  (or perhaps in the subgroup generated
by the cyclotomic units) is  (p - 1)/2 .  One can check that this is correct for  p = 11  but fails for  p = 13 .  In that case there is a
relation between these units and hence they do not generate a subgroup
of the correct rank.

        For a prime  p  the problem is easily reduced to a difficult problem
in number theory via the following lemma.


Lemma.  Let  p  be a prime,  G  a cyclic group of order  p  with generator
t ,  and  ζ  a primitive p-th root of unity.  The map  ZG → Z[ζ]  given
by  t → ζ  induces an isomorphism of  U(ZG)  with the subgroup of
U(Z[ζ])  consisting of all units congruent to  ±1  modulo  ζ - 1 .  In
particular, the image has index  (p - 1)/2  in the group  U(Z[ζ]) .


        Note that Milnor [52, Lemma 12.10, p.408] (see [3]) has shown how
to find a subgroup of  F  of finite index.

        H. Bass pointed out that in case  G  is a finite group (not neces-
sarily abelian) the group  U(ZG)  is an arithmetic group and hence by a
deep result of Borel and Harish-Chandra it is finitely presented.


Problem 16.  Let  G  be a finite group.  Find generators and relations
for  U(ZG) .


        One can obtain some information by abelianizing.  By Milnor's
theorem, one sees that  U(ZG)$^{ab}$  maps to a subgroup of finite index in

$K_1(ZG)$ . Thus the rank of $U(ZG)^{ab}$ is greater than or equal to the rank of $K_1(ZG)$ which by a theorem of Bass [3, Corollary 6.3] has rank $r - q$ where $r$ is the number of irreducible real representations and $q$ is the number of irreducible rational representations of $G$ . By Section 5, we have the decomposition

$$U(ZG)^{ab} = U(Z) \times G^{ab} \times F \ .$$

Problem 17. Determine the structure of the finitely generated abelian group $F$ . What is its rank? Compute the kernel of the map $F \to K_1(ZG)$ .

Problem 18. Find a set of generators for $F$ . Find a set of generators for $K_1(ZG)$ modulo torsion. In particular, is $F \to K_1(ZG)$ surjective modulo torsion?

Theorem. (Rothaus). [72]. Let $D_{10}$ be the dihedral group with generator $g,h$ and relations $g^5 = h^2 = 1$ , $hgh = g^{-1}$ . Then the element $u = (-1 + g - g^2 + g^3 + g^4) + h(1 - 2g + g^2)$ is a unit of $ZD_{10}$ with inverse the element $(1 - g - g^2 + g^3 + g^4) + h(-2g + g^3 + g^4)$ . Furthermore, the image of $u$ generates a free summand of $K_1(ZD_{10})$ and

$$K_1(ZD_{10}) = U(Z) \times D_{10}^{ab} \times (\bar{u}) \ .$$

With a little work one can use the results of Hughes and Pearson (see Section 8 above) to solve Problems 16, 17 and 18 for the group $S_3$ .

Using their map $U(ZS_3) \to GL_2(Z)$ one sees that the image $I$ has index
4 . The congruence subgroup $C$ consisting of those matrices of $GL_2(Z)$
congruent to 1 modulo 3 is a subgroup of $I$ of index $48/4 = 12$ .
Thus one easily checks that $I/C$ is isomorphic to $U(Z) \times S_3$ . Now $C$
is a subgroup of $SL_2(Z)$ of index 24 and hence its Euler characteristic
is $-1/12 \cdot 24 = -2$ . As $SL_2(Z)$ is the amalgamated free product of two
finite groups, it follows that any torsion-free subgroup is free. Thus
$C$ is a free group of rank $1 - (-2) = 3$ . One thus obtains that $U(ZS_3)$
is the direct product of $U(Z)$ with a group which is the semidirect
product of a free group on three generators and $S_3$ . It thus only remains
to determine the action of $S_3$ on the free group in order to obtain the
presentation. (I would like to thank K. Brown for pointing out that $C$ is
a free group on three generators.) Using standard techniques from the
theory of the modular group one can show that the group $C$ is generated
by the matrices

$$A = \begin{pmatrix} 1 & 3 \\ 0 & 1 \end{pmatrix} , \quad B = \begin{pmatrix} 1 & 0 \\ -3 & 1 \end{pmatrix} , \quad C = \begin{pmatrix} 4 & -3 \\ 3 & -2 \end{pmatrix} .$$

The group $S_3$ is generated by the matrices

$$a = \begin{pmatrix} 1 & -1 \\ 0 & -1 \end{pmatrix} , \quad b = \begin{pmatrix} 0 & -1 \\ 1 & -1 \end{pmatrix} .$$

Thus the group $V(ZS_3)$ has the presentation

$$\langle a,b,A,B,C \mid a^2,b^3,abab,aAaA,aBaC,bAb^{-1}B^{-1},bAb^{-1}C^{-1},bCb^{-1}A^{-1} \rangle .$$

Now $V(ZS_3)^{ab} \simeq <a,A \mid aAa^{-1}A^{-1},a^2,A^2>$ is the Klein four group. Thus F is cyclic of order 2 and does not inject into $K_1(ZS_3)$ .

Let $D_8 = <\alpha,\beta \mid \alpha^4,\beta^2,\alpha\beta\alpha\beta>$ . A similar argument shows that the homomorphism $ZD_8 \to M_2(Z)$ given by

$$\alpha \to a = \begin{pmatrix} 0 & 1 \\ -1 & 0 \end{pmatrix} , \quad \beta \to b = \begin{pmatrix} 0 & 1 \\ 1 & 0 \end{pmatrix}$$

is injective on units. The image of the units contains the congruence subgroup of $GL_2(Z)$ modulo 4 with index 2 . It is torsion-free and hence free of rank 3 with generators

$$A = \begin{pmatrix} 1 & 0 \\ 4 & 1 \end{pmatrix} , \quad B = \begin{pmatrix} 1 & 4 \\ 0 & 1 \end{pmatrix} , \quad C = \begin{pmatrix} 5 & 2 \\ 2 & 1 \end{pmatrix} .$$

One obtains the following presentation for $V(ZD_8)$ :

$$<a,b,A,B,C \mid a^4,b^2,aAa^{-1}B,aBa^{-1}A,aCa^{-1}C,baba,bAbB^{-1},bBbA^{-1},bCbB^{-1}CA^{-1}> .$$

Also $V(ZD_8)^{ab}$ is an elementary abelian 2-group of rank 4 with generators the images of $a,b,A,C$ . F is an elementary abelian 2-group of rank 2 with generators the images of A,C . F maps to 1 in $K_1(ZD_8)$ .


§ 10. **A Problem of Rothaus.** Let G be an arbitrary group. Then E in $GL_n(ZG)$ is called **positive** if $[\det \varepsilon(E)]^k \det \rho(E) > 0$ for all proper orthogonal representations $\rho : G \to SO(k)$ . Recall that $K_1(ZG) = GL(ZG)^{ab}$

and  $Wh(G) = K_1(ZG)/image$ of $\pm G$ .  Define  $Ro(G)$  to be  $Wh(G)$  modulo
the subgroup of all positive elements.  As any square is positive,  $Ro(G)$
is an elementary 2-group.

<u>Problem 19</u>.  Compute the rank of  $Ro(G)$ .  In particular, characterize
those groups  $G$  for which  $rank\ Ro(G) > 0$ .

Rothaus [72] has shown that  $rank\ Ro(G) = 0$  if  $G$  is finite abelian
and that  $rank\ Ro(D_{10}) = 1$ .  The element  $u$  mentioned in Section 9 above
maps to the generator of  $Ro(D_{10})$ .

If  $rank\ Ro(G) > 0$ , one can show that certain group extensions of
$G$  given by generators and relations are nontrivial [72].  These results
have topological applications via the work of Cohen [20].

§<u>11</u>.  <u>Miscellaneous</u>.  Coleman and Passman [23] have shown that if  $F$  is
a field of characteristic  $p$  and  $G$  is a finite non-abelian p-group,
then some subgroup of  $U(FG)$  will have  $Z_p \wr Z_p$  (wreath product) as a
quotient.

Various computations of  $U(RG)$  have been given.  See [68] and
[69], for example.

# REFERENCES

1.  Ayoub, R. G. and C. Ayoub.  "On the group ring of a finite abelian group," Bull. Austral. Math. Soc. 1(1969), 245-261.  MR40#5746.

2.  Bachmuth, S., E. Formanek and H. Y. Mochizuki.  "IA-automorphisms of certain two-generator torsion-free groups," to appear.

3.  Bass, H.  "The Dirichlet unit theorem, induced characters and White-head groups of finite groups," Topology 4(1966), 391-410.  MR33#1341.

4.  _____.  Algebraic K-Theory, Benjamin, New York, 1968.  MR40#2736.

5.  _____.  "Euler characteristics and characters of discrete groups," to appear.

6.  Bateman, J. M.  Finite Group Algebras with Solvable Unit Groups, Thesis, Vanderbilt University, 1967.  (See Dissert. Abstr. B28, No. 10, 4191(1968).)

7.  _____.  "On the solvability of unit groups of group algebras," Trans. Amer. Math. Soc. 157(1971), 73-86.  MR43#2118.

8.  Bateman, J. M. and D. B. Coleman.  "Group algebras with nilpotent unit groups," Proc. Amer. Math. Soc. 19(1968), 448-449.  MR36#5238.

9.  Berman, S. D.  "On certain properties of integral group rings," Dokl. Akad. Nauk SSSR (N.S.) 91(1953), 7-9.  MR15,99.

10. _____.  "On the equation $x^m = 1$ in an integral group ring," Ukrain. Mat. Ž. 7(1955), 253-261.  MR17,1048.

11. Berman, S. D. and A. R. Rossa.  "Integral group rings of finite and periodic groups," Algebra and Math. Logic:  Studies in Algebra, Izdat. Kiev Univ., Kiev, 1966, 44-53.  MR35#265.

12. _____. "The Sylow p-subgroup of a group
    algebra over a countable abelian p-group," <u>Dopovīdī</u> <u>Akad</u>. <u>Nauk</u>
    <u>Ukraïn</u>. RSR Ser. A, 1968, 870-872. MR40#1479.

13. Bhattacharya, P. B. and S. K. Jain. "A note on the adjoint group
    of a ring," <u>Arch</u>. <u>der</u> <u>Math</u>. 21(1970), 366-368. MR42#7705.

14. Bovdi, A. A. "Group rings of torsion free groups," <u>Sibirsk</u>. <u>Mat</u>.
    <u>Ž</u>. 1(1960), 555-558. MR24#A773.

15. _____. "Periodic normal divisors of the multiplicative
    group of a group ring, I," <u>Sibirsk</u>. <u>Mat</u>. <u>Ž</u>. 9(1968), 495-498 =
    <u>Siberian</u> <u>Math</u>. <u>J</u>. 9(1968), 374-376. MR37#2853.

16. _____. "Periodic normal divisors of the multiplicative
    group of a group ring, II," <u>Sibirsk</u>. <u>Mat</u>. <u>Ž</u>. 11(1970), 492-511 =
    <u>Siberian</u> <u>Math</u>. <u>J</u>. 11(1970), 374-388. MR43#4930.

17. _____. <u>Group</u> <u>Rings</u>, Uzgorod, 1974.

18. Bovdi, A. A. and I. I. Hripta. "Normal subgroups of a multiplica-
    tive group of a ring," <u>Mat</u>. <u>Sb</u>. 87(1972), 338-350 = <u>Math</u>. <u>USSR-Sb</u>.
    16(1972), 349-362. MR46#200.

19. Cohen, M. M. <u>A</u> <u>Course</u> <u>in</u> <u>Simple-Homotopy</u> <u>Theory</u>, Springer-Verlag,
    Berlin and New York, 1973.

20. _____. "Whitehead torsion, group extensions, and Zeeman's
    conjecture in high dimensions," to appear.

21. Cohn, J. A. and D. Livingstone. "On the structure of group alge-
    bras, I," <u>Canad</u>. <u>J</u>. <u>Math</u>. 17(1965), 583-593. MR31#3514.

22. Coleman, D. B. "On the modular group ring of a p-group," <u>Proc</u>.
    <u>Amer</u>. <u>Math</u>. <u>Soc</u>. 15(1964), 511-514. MR29#2306.

23.  Coleman, D. B. and D. S. Passman.  "Units in modular group
     rings," Proc. Amer. Math. Soc. 25(1970), 510-512.  MR41#6968.

24.  Curtis, C. W. and I. Reiner.  Representation Theory of Finite Groups
     and Associative Algebras, Wiley, New York, 1962.  MR26#2519.

25.  Dade, E. C.  "Deux groupes finis distincts ayant la même algèbre de
     group sur tout corps," Math. Z. 119(1971), 345-348.  MR43#6329.

26.  Dennis, R. K.  "Units of group rings," J. Algebra, to appear.

27.  Eldridge, K. E.  "On ring structures determined by groups," Proc.
     Amer. Math. Soc. 23(1969), 472-477.  (See correction Ibid. 25(1970),
     202.)  MR39#6923 (41#1797).

28.  _____.  "On Rings and Groups of Units," Rings, Modules and
     Radicals, Colloq. Math. Soc. János Bolyai 6(1973), 177-191.  Z262.160ɛ

29.  _____.  "On normal subgroups of modular group algebras,"
     Notices Amer. Math. Soc. 17(1970), 764, No. 677-16-2.

30.  Farrell, F. T. and W. C. Hsiang.  "A formula for $K_1 R_\alpha [T]$ ," Applica-
     tions of Categorical Algebra, pp. 192-218, Proc. Symp. Pure Math. 17,
     Amer. Math. Soc., Providence, 1970.  MR41#5457.

31.  Fisher, J. L., M. M. Parmenter and S. K. Sehgal.  "Group rings with
     solvable n-Engel unit groups," Proc. Amer. Math. Soc., to appear.

32.  Formanek, E.  Matrix Techniques in Polycyclic Groups, Thesis, Rice
     University, 1970.

33.  Grover, J. and E. Malek.  "The ranks of Whitehead groups," preprint.

34.  Herstein, I. N.  Notes from a Ring Theory Conference, CBMS No. 9,
     Amer. Math. Soc., Providence, 1971.  MR47#1840.

35.  Higman, G.  "The units of group rings," _Proc_. _London_ _Math_. _Soc_. (2)
     46(1940), 231-248.  MR2,5.

36.  Hripta, I. I.  _On_ _the_ _Multiplicative_ _Group_ _of_ _a_ _Group_ _Ring_, Thesis,
     Uzgorod, 1971.

37.  _____.  "The nilpotence of the multiplicative group of a group
     ring," _Mat_. _Zametki_ 11(1972), 191-200 = _Math_. _Notes_ 11(1972), 119-
     124.  MR47#312.

38.  _____.  "On nilpotency of the multiplicative group of a group
     ring," _Latvian_ _Mathematical_ _Yearbook_ 13, 119-127.  Izdat. "Zinatne,"
     Riga, 1973.  MR49#5077.

39.  Hughes, I. and K. R. Pearson.  "The group of units of the integral
     group ring $ZS_3$," _Canad_. _Math_. _Bull_. 15(1972), 529-534.  MR48#4089.

40.  Hughes, I. and C. Wei.  "Group rings with only trivial units of
     finite order," _Canad_. _J_. _Math_. 24(1972), 1137-1138.  MR47#313.

41.  Jackson, D. A.  "The groups of units of the integral group rings of
     finite metabelian and finite nilpotent groups," _Quart_. _J_. _Math_.
     _Oxford_ _Ser_. (2)20(1969), 319-331.  MR40#2766.

42.  Jacobson, N.  "Some remarks on one-sided inverses," _Proc_. _Amer_. _Math_.
     _Soc_. 1(1950), 352-355.  MR12,75.

43.  Jennings, S. A.  "The structure of the group ring of a p-group over
     a modular field," _Trans_. _Amer_. _Math_. _Soc_. 50(1941), 175-185.  MR3,34.

44.  Johnson, D. L.  "The modular group-ring of a finite p-group," _Proc_.
     _Amer_. _Math_. _Soc_., to appear.  Z264.20020.

45.  Kaplansky, I.  _Fields_ _and_ _Rings_, Chicago Lectures in Mathematics,
     Univ. Chicago Press, Chicago, 1969.  MR42#4345.

46. _____. "Problems in the theory of rings, revisited," Amer.
    Math. Monthly 77(1970), 445-454.   MR41#3510.

47. Lanski, C.  "Some remarks on rings with solvable units," Ring Theory
    R. Gordon, ed., Academic Press, New York, 1972, 235-240.  MR49#7311.

48. Lombardo-Radice, L.  "Intorno alle algebre legate ai gruppi di ordine
    finito," Rend. Semin. Mat. Roma (4)2(1938), 312-322.  Z20,341.

49. Losey, G.  "Are one-sided inverses two-sided inverses in a matrix
    ring over a group ring?"  Canad. Math. Bull. 13(1970), 475-479.
    MR42#7703 (44,p.1631).

50. _____.  "A remark on the units of finite order in the group ring
    of a finite group,"  Canad. Math. Bull. 17(1974), 129-130.  MR50#4719

51. Malek, E.  "On the group of units of a finite R-algebra," J. Algebra
    23(1972), 538-552.  MR47#207.

52. Milnor, J.  "Whitehead torsion," Bull. Amer. Math. Soc. 72(1966),
    358-426.  MR33#4922.

53. Montgomery, M. S.  "Left and right inverses in group algebras," Bull.
    Amer. Math. Soc. 75(1969), 539-540.  MR39#327.

54. Motose, K. and Y. Ninomiya.  "On the solvability of unit groups of
    group rings," Math. J. Okayama Univ. 15(1972), 209-214.  MR48#397.

55. Motose, K. and H. Tominaga.  "Group rings with nilpotent unit groups,"
    Math. J. Okayama Univ. 14(1969), 43-46.  MR41#1899.

56. _____.  "Group rings with solvable unit groups,"
    Math. J. Okayama Univ. 15(1971), 37-40.  MR46#5423.

57. Nakajima, A. and H. Tominaga.  "A note on group rings of p-groups,"
    Math. J. Okayama Univ. 13(1968), 107-109.  MR39#328.

58.  Obayashi, T.  "On integral group rings," (Japanese), Sûgaku 19
     (1967), 82-94.  MR37#4178.

59.  Olum, P.  "Self-equivalences of pseudo-projective planes," Topology
     4(1965), 109-127.  MR31#2725.

60.  _____.  "Self-equivalences of pseudo-projective planes, II,"
     Topology 10(1971), 257-260.  MR43#1185.

61.  Passman, D. S.  "Isomorphic groups and group rings," Pacific J. Math.
     15(1965), 561-583.  (See correction in Ibid. 43(1972), 823-824.)
     MR33#1381.

62.  _____.  Infinite Group Rings, Marcel Dekker, New York, 1971,
     pp. 110-114.  MR47#3500.

63.  _____.  "Advances in group rings," Israel J. Math. 19(1974),
     67-107.  MR50#9945.

64.  Pearson, K. R.  "On the units of a modular group ring," Bull.
     Austral. Math. Soc. 7(1972), 169-182.  MR48#2237.

65.  _____.  "On the units of a modular group ring, II," Bull.
     Austral. Math. Soc. 8(1973), 435-442.  MR49#10730.

66.  Peterson, R. D.  "One-sided inverses in rings," Canad. J. Math.
     27(1975), 218-224.  Z268.16005.

67.  Polcino Milies, C.  "Integral group rings with nilpotent unit groups,"
     to appear.

68.  Raggi Cárdenas, F. F.  "Units in group rings, I," An. Inst. Mat.
     Univ. Nac. Autónoma México 7(1967), 27-35.  MR38#2225.

69.  _____.  "Units in group rings, II," An. Inst. Mat.
     Univ. Nac. Autónoma México 8(1968), 91-103.  MR40#1498.

70. Rossa, A. R. "Group rings of p-groups over the ring of integral p-adic numbers," _Sibirsk_. _Math_. Ž. 9(1968), 220-222. MR36#6514.

71. Rossa, A. R. and S. D. Berman. "On integral group rings," _Third Scientific Conference of Young Mathematicians of the Ukraine_, Kiev, 1966, p.75.

72. Rothaus, O. S. "On the non-triviality of some group extensions given by generators and relations," _Bull_. _Amer_. _Math_. _Soc_. 82(1976), 284-28

73. Rudin, W. and H. Schneider. "Idempotents in group rings," _Duke_ _Math_. _J_. 31(1964), 585-602. MR29#5119.

74. Saksonov, A. I. "Certain integer-valued rings associated with a finite group," _Dokl_. _Akad_. _Nauk_ _SSSR_ 111(1966), 529-532. MR34#7676.

75. _____. "On group rings of finite p-groups over certain integral domains," _Dokl_. _Akad_. _Nauk_ _BSSR_ 11(1967), 204-207. MR35#27C

76. _____. "On group rings of finite groups, I." _Publ_. _Math_. _Debrecen_ 18(1971), 187-209. MR46#5425.

77. Sehgal, S. K. "On isomorphisms of integral group rings, I," _Canad_. _J_. _Math_. 21(1969), 410-413. MR41#366.

78. _____. "Units in commutative integral group rings," _Math_. _J_. _Okayama_ _Univ_. 14(1970), 135-138. MR44#6861.

79. _____. "Isomorphism of p-adic group rings," _J_. _Number_ _Theory_ 2(1970), 500-508. MR42#1917.

80. _____. "Lie properties in modular group algebras," _Lecture Notes in Math_., Springer-Verlag, Berlin and New York, 1973, pp. 152-160. MR49#5151.

81. _____. "Nilpotent elements in group rings," Manuscripta Math.
    15(1975), 65-80.  Z302.16010.

82. _____. "Certain algebraic elements in group rings," Archiv
    der Math. 26(1975), 139-143.

83. Sehgal, S. K. and H. J. Zassenhaus. "Integral group rings with nil-
    potent unit groups," to appear.

84. Shepherdson, J. C. "Inverses and zero divisors in matrix rings,"
    Proc. London Math. Soc. (3)1(1951), 71-85.  MR13,7.

85. Sinha, I. "Semi-simplicity of group rings with trivial units,"
    Tamkang J. Math. 5(1974), 107-108.  MR50#9946.

86. Stallings, J. "Whitehead torsion of free products," Ann. of Math.
    (2)82(1965), 354-363.  MR31#3518.

87. Swan, R. G. Algebraic K-Theory, Lecture Notes in Math. 76,
    Springer-Verlag, Berlin and New York, 1968.  MR39#6940.

88. _____. K-Theory of Finite Groups and Orders, Lecture Notes in
    Math. 149, Springer-Verlag, Berlin and New York, 1970.  MR46#7310.

89. Takahashi, S. "Some properties of the group ring over rational inte-
    gers of a finite group," Notices Amer. Math. Soc. 12(1965), 463,
    No. 65T-226.

90. Taussky, O. "Matrices of rational integers," Bull. Amer. Math. Soc.
    66(1960), 327-345.  MR22#10994.

91. Wall, C. T. C. "Norms of units in group rings," Proc. London Math.
    Soc. (3)29(1974), 593-632.  Z302.16013.

92. Zaitseva, M. I. "Right-orderable groups," Uch. Zap. Shuisk. Gos.
    Ped. Inst. 6(1958), 205-206.

93.  Zalesskii, A. E.   "Solvable subgroups of the multiplicative group
     of a locally finite algebra," Mat. Sb. 61(1963), 408-417.  MR26#6255

94.  Zalesskii, A. E. and A. V. Mikhalev.  "Group rings," Itogi Nauki
     i Tekhniki (Sobremennye Problemy Matematiki), Vol. 2, 1973, pp.5-118
     J. Soviet Math. 4(1975), 1-78.  Z293.16013.

95.  Zassenhaus, H.  "On the torsion units of finite group rings," Studies
     in Mathematics (in honor of A. Almeida Costa), Instituo de Alta
     Cultura, Lisbon, 1974, pp. 119-126.

96.  Zhmud', E. M. and G. C. Kurennoi.  "On finite groups of units of an
     integral group ring," Vesnik Har'kov. Gos. Univ. 26(1967), 20-26.
     MR40#2770.

(The entries above are followed by a reference to a review from either
the Mathematical Reviews or Zentralblatt für Mathematik.)

POWER-CANCELLATION OF MODULES

K. R. Goodearl

University of Utah

§1.  **Introduction**.  This paper is concerned with conditions which
ensure that even though a module  A  may not necessarily cancel from
a direct sum  $A \oplus B \cong A \oplus C$ , it can at least be concluded that
$B^n \cong C^n$  for some positive integer  n .  This conclusion is obtained
from a type of stable range condition on the endomorphism ring of  A ,
which holds, for example, when  A  is a finitely generated module over
any subring of a finite dimensional algebra over the rationals.

§2.  **Stable Range and Cancellation**.  Since our power-cancellation
results are modelled directly on the cancellation results obtained
from the stable range conditions of algebraic K-theory, we present a
discussion of this material in the present section.

   A ring  R  is said to have  1  **in the stable range** provided that
whenever  $ax + b = 1$  in  R , there exists  y  in  R  such that  $a + by$
is a unit in  R .  [The first integer  1  in this definition refers to
the reduction of a "unimodular row of length  2 ," namely the pair
(a,b) , to a "unimodular row of length  1 ," namely the element

a + by . Higher stable range conditions are also studied, where  " n
in the stable range" denotes the ability to reduce a "unimodular row"
from length  n + 1  to length  n . References on algebraic K-theory
such as [8] may be consulted for the precise definitions and their
consequences.] Although this definition is asymmetric (the coefficients
x  and  y  occur on the right of  a  and  b ), Theorem 2 shows that the
property involved is, in fact, left-right symmetric.

For example, every Artinian ring has  1  in the stable range [8,
Lemma 11.8]. Since it is trivial to check that this property lifts
modulo the Jacobson radical, it follows that any ring  R  for which
$R/J(R)$  is Artinian must have  1  in the stable range. If  R  is any
commutative Noetherian ring with Krull dimension zero, then any R-
algebra which is finitely generated as an R-module has  1  in the
stable range [8, Theorem 12.3]. More generally, if  R  is a commutative
ring such that  $R/J(R)$  is von Neumann regular, and if  S  is any R-
algebra which is (locally) finitely generated as an R-module, then the
endomorphism ring of every finitely generated S-module has  1  in the
stable range [4, Theorem 18]. These examples are all in some sense
"zero-dimensional." In general, rings which are in some sense "one-
dimensional" or higher usually do not have  1  in the stable range.
For example, the ring of integers  Z  does not have  1  in the stable
range;  $2 \cdot 3 - 5 = 1$ , but  $2 - 5y$  is never a unit in  Z .

Various cancellation theorems in algebraic K-theory have been
proved by means of the stable range conditions, but the first explicit
proof that stable range  1  (with no other hypotheses) implies

cancellation seems to be the following theorem of Evans.

Theorem 1.  [1, Theorem 2].  Let  A  be a module whose endomorphism
ring has  1  in the stable range.  If  B  and  C  are any modules such
that  $A \oplus B \cong A \oplus C$ ,  then  $B \cong C$ .

The proof of Theorem 1 actually reveals a stronger result.  Start-
ing with a module  M  with decompositions  $M = A_1 \oplus B_1 = A_2 \oplus B_2$  (where
$A_1 \cong A_2 \cong A$ ,  $B_1 \cong B$ ,  $B_2 \cong C$ ),  the stable range condition on
End(A)  is used to find a common complement for  $B_1$  and  $B_2$ ,  i.e.,
$M = D \oplus B_1 = D \oplus B_2$  for some  D .  (For details see the proof of
Theorem 2.)  This substitution property is equivalent to stable range
1 ,  as proved by Fuchs for quasi-projective modules with projective
covers [2, Theorem 3], and in general by Warfield [10, Theorem 1].  A
slight addition to Warfield's proof yields, in addition, the categorical
dual of the substitution property, along with the left-right symmetry
of stable range  1 ,  as follows.

Theorem 2.  Let  A  be a right R-module, and set  $E = \text{End}_R(A)$ .  Then
the following conditions are equivalent:
  (a)  If  $ax + b = 1$  in  E ,  then there exists  y  in  E  such
       that  $a + by$  is a unit.
  (b)  If  $xa + b = 1$  in  E ,  then there exists  y  in  E  such
       that  $a + yb$  is a unit.
  (c)  Given any right R-module decomposition  $M = A_1 \oplus B_1 = A_2 \oplus B_2$

with each $A_i \cong A$ , there exists a submodule $C \leqq M$ such

that $M = C \oplus B_1 = C \oplus B_2$ .

(d)  Given any right R-module decomposition $M = A_1 \oplus B_1 = A_2 \oplus B_2$

with each $A_i \cong A$ , there exists a submodule $D \leqq M$ such that

$M = A_1 \oplus D = A_2 \oplus D$ .


Proof.  In order to organize the proof efficiently, we require an

additional condition (d'):  Given $M = A_1 \oplus B_1 = A_2 \oplus B_2$ with $A_1 \cong$

$A_2 \cong B_1 \cong A$ , there exists $D \leqq M$ such that $M = A_1 \oplus D = A_2 \oplus D$ .

(a) $\Rightarrow$ (c):  Using the decomposition $M = A_1 \oplus B_1 \cong A \oplus B_1$ , we

obtain projections $p_1 : M \to A$ , $p_2 : M \to B_1$ and injections $q_1 : A \to M$

$q_2 : B_1 \to M$ such that $p_1 q_1 = 1_A$ , $p_i q_j = 0$ for $i \neq j$ ,

$q_1 p_1 + q_2 p_2 = 1_M$ , and $\ker p_1 = B_1$ .  Using the decomposition $M =$

$A_2 \oplus B_2 \cong A \oplus B_2$ , we obtain a projection $f : M \to A$ and an injection

$g : A \to M$ such that $fg = 1_A$ and $\ker f = B_2$ .

Now $1_A = f(q_1 p_1 + q_2 p_2)g = (fq_1)(p_1 g) + (fq_2 p_2 g)$ .  Setting

$a = fq_1$ , $x = p_1 g$ , and $b = fq_2 p_2 g$ , we have $a$ , $x$ , $b$ in $E$ such

that $ax + b = 1$ .  By (a), there exists $y$ in $E$ such that $a + by$

is a unit.

Set $k = q_1 + q_2 p_2 gy : A \to M$ and $C = k(A)$ .  Since $fk = a + by$

is an isomorphism, we infer that $M = k(A) \oplus (\ker f) = C \oplus B_2$ .  Since

$p_1 k = p_1 q_1 = 1_A$ , we likewise obtain $M = k(A) \oplus (\ker p_1) = C \oplus B_1$ .

(c) $\Rightarrow$ (d'):  By (c), there exists $C \leqq M$ such that $M = C \oplus B_1 =$

$C \oplus B_2$ , whence $B_1 \cong B_2$ .  Since $B_1 \cong A$ , another application of (c)

gives us $D \leqq M$ such that $M = A_1 \oplus D = A_2 \oplus D$ .

(d') $\Rightarrow$ (b):  Set  $M = A \oplus A$ , and let  $p_1, p_2 : M \to A$ ,
$q_1, q_2 : A \to M$  be the canonical projections and injections of this
direct sum.  Set  $A_1 = q_1(A)$  and  $B_1 = q_2(A)$ , so that  $M = A_1 \oplus B_1$
with  $A_1 \cong B_1 \cong A$ .  Set  $f = xp_1 + p_2 : M \to A$  and  $g = q_1 a + q_2 b$ :
$A \to M$ .  Then  $fg = xa + b = 1_A$ , whence  $M = g(A) \oplus (\ker f)$ .  Set
$A_2 = g(A)$  and  $B_2 = \ker f$ , so that  $M = A_2 \oplus B_2$  with  $A_2 \cong A$ .
According to (d'), there exists  $D \leq M$  such that  $M = A_1 \oplus D = A_2 \oplus D$ .

Since  $A_1 \cong A$ , we now have an epimorphism  $h : M \to A$  such that
$\ker h = D$ .  Also, we have a monomorphism  $g : A \to M$  such that
$M = g(A) \oplus (\ker h)$ , from which we infer that  $hg$  is an isomorphism.
Likewise, we see from the decomposition  $M = q_1(A) \oplus (\ker h)$  that
$hq_1$  is an isomorphism.  Inasmuch as  $hq_1 a + hq_2 b = hg$  is an isomorphism,
so is  $a + (hq_1)^{-1} hq_2 b$ .  Thus we have  $y = (hq_1)^{-1} hq_2$  in  $E$  such that
$a + yb$  is a unit.

(b) $\Rightarrow$ (a):  The implications above show that (a) $\Rightarrow$ (b) for any
ring  $E$ .  By symmetry, (b) $\Rightarrow$ (a).

(d) $\Rightarrow$ (d') a priori.

(b) $\Rightarrow$ (d):  Let  $p_1 : M \to A$ ,  $p_2 : M \to B_1$  and  $q_1 : A \to M$ ,
$q_2 : B_1 \to M$  be the projections and injections from the decomposition
$M = A_1 \oplus B_1 \cong A \oplus B_1$ .  Let  $f : M \to A$  and  $g : A \to M$  be the projection and injection from the decomposition  $M = A_2 \oplus B_2 \cong A \oplus B_2$ .

Setting  $x = fq_1$ ,  $a = p_1 g$  and  $b = fq_2 p_2 g$ , we have  $x$ , $a$ , $b$
in  $E$  such that  $xa + b = 1$ .  By (b), there exists  $y$  in  $E$  such that
$a + yb$  is a unit.  Now set  $k = p_1 + yfq_2 p_2 : M \to A$  and  $D = \ker k$ .
Since  $kg = a + yb$  and  $kq_1 = 1_A$  are isomorphisms, we conclude that

$M = g(A) \oplus (\ker k) = A_2 \oplus D$  and  $M = q_1(A) \oplus (\ker k) = A_1 \oplus D$ .

§3.  **Power-substitution and Power-cancellation.**  The results of
Section 1 may be summarized by saying that stable range 1 implies
substitution, which in turn implies cancellation. In this section
we derive, by direct analogy, a property which we call **power-cancellation:**
namely, that $A \oplus B \cong A \oplus C$ implies $B^n \cong C^n$ for some positive integer
n . We obtain power-cancellation from a type of substitution condition,
which in turn is equivalent to a type of stable range condition on
$\text{End}(A)$ . All of this is formal categorical manipulation, and the
proofs are obvious generalizations of the techniques of Theorem 2.
These results obtain substance only because of the large class of
rings which satisfy these conditions (a much larger class than those
rings which have 1 in the stable range).

We use $M_n(R)$ to denote the ring of all $n \times n$ matrices over a
ring R , and we use I to denote the identity matrix in any $M_n(R)$ .
Given r in R and P in $M_n(R)$ , we use rP to denote the matrix
obtained from P by multiplying each entry on the left by r .

**Theorem 3.** [3, Theorem 2.1]. Let A be a right R-module, and set
$E = \text{End}_R(A)$ . Then the following conditions are equivalent:

    (a) If $ax + b = 1$ in E , then there exist a positive integer
        n and a matrix Q in $M_n(E)$ such that $aI + bQ$ is a unit

in $M_n(E)$ .

(b)  Given any right R-module decomposition  $M = A_1 \oplus B_1 = A_2 \oplus B_2$

with each  $A_i \cong A$ , there exist a positive integer  $n$  and a

submodule  $C \leq M^n$  such that  $M^n = C \oplus B_1^n = C \oplus B_2^n$ .

Proof.  For each positive integer  $n$ , there is an additive functor from

Mod-R $\to$ Mod-R  which carries any module  $D$  to  $D^n$ .  For any map

$f : D \to E$ , we use  $f^*$  to denote the image of  $f$  under this functor.

Thinking of  $f^*$  as an  $n \times n$  matrix with entries from  $\text{Hom}_R(D,E)$ ,

$f^*$  is a diagonal matrix with all diagonal entries equal to  $f$ .

(a) $\Rightarrow$ (b):  Using the decomposition $M = A_1 \oplus B_1 \cong A \oplus B_1$ , we

obtain projections  $p_1 : M \to A$ ,  $p_2 : M \to B_1$  and injections

$q_1 : A \to M$ ,  $q_2 : B_1 \to M$ .  Using the decomposition  $M = A_2 \oplus B_2 \cong A \oplus B_2$ ,

we obtain a projection  $f : M \to A$  and an injection  $g : A \to M$ .

Since  $1_A = f(q_1 p_1 + q_2 p_2)g$ , we obtain  $ax + b = 1$  in  $E$ , where

$a = fq_1$ ,  $x = p_1 g$  and  $b = fq_2 p_2 g$ .  According to (a), there exist

$n > 0$  and  $Q$  in  $M_n(E)$  such that  $aI + bQ$  is a unit in  $M_n(E)$ .

Identifying  $M_n(E)$  with  $\text{End}_R(A^n)$  in the obvious manner, we thus

obtain a map  $h = Q : A^n \to A^n$  such that  $f^*q_1^* + f^*q_2^*p_2^*g^*h = a^* + b^*h$

is an automorphism of  $A^n$ .

Set  $k = q_1^* + q_2^*p_2^*g^*h : A^n \to M^n$  and  $C = k(A^n)$ .  Since  $f^*k$

is an isomorphism, we find that  $M^n = k(A^n) \oplus (\ker f^*) = C \oplus B_2^n$ .

Since  $p_1^*k = p_1^*q_1^*$  is the identity map on  $A^n$ , we also have

$M^n = k(A^n) \oplus (\ker p_1^*) = C \oplus B_1^n$ .

(b) $\Rightarrow$ (a): Set $M = A \oplus A$, and let $p_1, p_2 : M \to A$, $q_1, q_2 : A \to M$ be the canonical projections and injections of this direct sum. Set $A_1 = q_1(A)$ and $B_1 = q_2(A)$, so that $M = A_1 \oplus B_1$ with $A_1 \cong A$. Define maps $f = ap_1 + bp_2 : M \to A$ and $g = q_1 x + q_2 : A \to M$. Observing that $fg = ax + b = 1_A$, we see that $M = g(A) \oplus (\ker f)$. Set $A_2 = g(A)$ and $B_2 = \ker f$, so that $M = A_2 \oplus B_2$ with $A_2 \cong A$.

According to (b), there exist $n > 0$ and $C \leq M^n$ such that $M^n = C \oplus B_1^n = C \oplus B_2^n$. Since $C$ and $A_1^n$ are both complements for $B_1^n$ in $M^n$, we see that $C \cong A_1^n \cong A^n$. As a result, there exists a monomorphism $h : A^n \to M^n$ such that $h(A^n) = C$. Inasmuch as $p_1^* : M^n \to A^n$ is an epimorphism and $M^n = C \oplus B_1^n = h(A^n) \oplus (\ker p_1^*)$, we infer that $p_1^* h$ is an isomorphism. Similarly, $f^* : M^n \to A^n$ is an epimorphism and $M^n = C \oplus B_2^n = h(A^n) \oplus (\ker f^*)$, whence $f^* h$ is an isomorphism. Since $f^* h = a^* p_1^* h + b^* p_2^* h$, we conclude that $a^* + b^* p_2^* h (p_1^* h)^{-1}$ is an automorphism of $A^n$.

Using the identification of $\operatorname{End}_R(A^n)$ with $M_n(E)$, we thus have $Q = p_2^* h (p_1^* h)^{-1}$ in $M_n(E)$ such that $aI + bQ = a^* + b^* Q$ is a unit in $M_n(E)$.

We say that a ring $E$ has the **right power-substitution property** if $E$ satisfies condition (a) of Theorem 3, or equivalently, if the right module $A = E_E$ satisfies condition (b) of Theorem 3. We do not know whether this condition is left-right symmetric.

Corollary 4. Let  A  be a right R-module such that  $End_R(A)$  has the right power-substitution property. If  B  and  C  are any right R-modules such that  $A \oplus B \cong A \oplus C$ , then  $B^n \cong C^n$  for some positive integer n .

Obviously any ring which has  1  in the stable range also satisfies right power-substitution, and there is a sense in which power-substitution and stable range  1  are nearly equivalent. Given a ring  R  and positive integers  $k|n$ , there is a natural ring map  $M_k(R) \to M_n(R)$ . Considering the positive integers as a directed set ordered by divisibility, we thus obtain a directed system of matrix rings over  R , and we can form the direct limit  $S = \lim_{\to} M_n(R)$ . It is clear from the definitions that  S  has  1  in the stable range if and only if every  $M_n(R)$  satisfies right power-substitution. Because it is not known whether power-substitution is preserved in matrix rings, we cannot say that  S  has stable range  1  if and only if  R  has right power-substitution. (If this were true, it would follow from Theorem 2 that power-substitution is left-right symmetric.)

In general, power-substitution is weaker than stable range  1 . For example, the integers  Z  have power-substitution (Corollary 7), but we have seen that  Z  does not have  1  in the stable range. More generally, every commutative ring which is integral over  Z  satisfies power-substitution [3, Corollary 3.13]. In particular, the ring of algebraic integers in any algebraic number field satisfies power-substitution. This might lead one to expect that power-substitution is a property of Dedekind domains, or perhaps at least of principal ideal domains. This is false, however, for the polynomial ring $F[x]$ over a field $F$ of characteristic zero never satisfies power-substitution [3, Corollary 3.8]. For non-commutative examples of power-substitution, we have any ring $R$ whose additive group has finite rank (Theorem 10).

For commutative rings, power-substitution is equivalent to a condition which is easier to check than the original definition, as follows.


Proposition 5. [3, Proposition 3.2]. A commutative ring $R$ satisfies the power-substitution property if and only if whenever $ax + b = 1$ in $R$, there exist a positive integer $n$ and an element $y$ in $R$ such that $a^n + by$ is a unit in $R$.


Proof. First assume that $R$ satisfies power-substitution. Given $ax + b = 1$ in $R$, there exist $n > 0$ and $Q$ in $M_n(R)$ such that $aI + bQ$ is a unit in $M_n(R)$. Then $\det(aI + bQ)$ is a unit in $R$, and we observe that $\det(aI + bQ) = a^n + by$ for some $y$ in $R$.

Conversely, let $ax + b = 1$ in $R$ and assume that $a^n + by$ is a unit in $R$, for some $n > 0$ and some $y$ in $R$. Now

$$a^n + by = a^n + by(ax + b)^{n-1} = a^n + a_1 a^{n-1}b + a_2 a^{n-2}b^2 + \cdots + a_n b^n$$

for suitable $a_1, \ldots, a_n$ in $R$. According to [3, Lemma 3.1], there exists $Q$ in $M_n(R)$ for which $\det(aI + bQ) = a^n + by$, whence $aI + bQ$ is a unit in $M_n(R)$.

Corollary 6. Let  R  be a commutative ring such that for all nonzero

b  in  R , the group of units of  R/bR  is torsion.  (In particular,

this holds if  R/bR  is finite for all nonzero  b  in  R .)  Then  R

satisfies power-substitution.


Proof.  Let  $ax + b = 1$  in  R .  If  $b = 0$ , then  $a^1 + b0$  is a unit

in  R , hence we may assume that  $b \neq 0$ .  Then we have an element  $\bar{a}$

in the group of units of  R/bR .  Since this group is torsion, we must

have  $\bar{a}^{-n} = 1$  for some  $n > 0$ , hence  $a^n + by = 1$  for some  y  in  R .


Corollary 7.  All subrings of the rationals satisfy power-substitution.


Corollary 8.  The ring  C(X)  of all continuous real-valued functions

on a compact Hausdorff space  X  satisfies power-substitution.


Proof.  If  $fg + h = 1$  in  C(X) , then the functions  f  and  h  are

not both zero at any point of  X .  As a result,  $f^2 + h^2$  takes on

only positive values and so is a unit in  C(X) .


Corollary 9.  If  F  is a field which is an algebraic extension of a

finite field, then the polynomial ring  F[x]  satisfies power-

substitution.


Proof.  Note that  F  is a directed union of finite subfields  $F_i$ , whence $F[x]$ is a directed union of the subrings $F_i[x]$. Each $F_i[x]$ satisfies power-substitution by Corollary 6, hence so does $F[x]$.

In fact, the algebraic extensions of finite fields are the only fields $F$ for which $F[x]$ satisfies power-substitution [3, Corollary 3.8].

An abelian group $A$ has **finite rank** [6, p.49] provided there exists a positive integer $n$ such that every finitely generated subgroup of $A$ can be generated by $n$ elements. Equivalently, $A$ has finite rank if and only if there is a positive integer $n$ such that $A/T(A)$ can be embedded in a vector space of dimension $n$ over $Q$ and the socle of each p-primary component of $T(A)$ has dimension at most $n$ over $Z/pZ$.

We refer to a ring $R$ whose additive group has finite rank as a **finite rank Z-algebra**. ( $Z$ denotes the rational integers.)

**Theorem 10.** [**3**, Corollary 4.13]. If $R$ is any direct limit of finite rank Z-algebras, then $R$ satisfies the right and left power-substitution properties.

The torsion-free commutative case of this result is relatively straightforward to prove [3, Theorem 3.10], and it is not hard to derive the general commutative case as well. The proof of the

noncommutative case, however, involves a complicated series of reduc-
tions to various special cases, and is in definite need of simplification.

## §4.  Applications.

Theorem 11.  [3, Theorem 5.3].  Let  S  be a commutative ring which is
integral over  Z , let  T  be an S-algebra which is finitely generated
as an S-module, and let  R  be any subring of  T .  Let  A  be a
finitely generated right R-module, and let  B  and  C  be any right
R-modules.  If  $A \oplus B \cong A \oplus C$ , then  $B^n \cong C^n$  for some positive integer
n .

Proof.  It is not hard to show that  $\operatorname{End}_R(A)$  is a direct limit of
finite rank Z-algebras [3, Lemma 5.2].  Then apply Theorem 10 and
Corollary 4.

Theorem 12.  [3, Theorem 5.4].  Let  R  be a right nonsingular ring
whose maximal right quotient ring is a direct limit of finite rank Z-
algebras.  Let  A  be a finite dimensional nonsingular right **R**-module,
and let  B  and  C  be any right R-modules.  If  $A \oplus B \cong A \oplus C$ , then
$B^n \cong C^n$  for some positive integer  n .

Theorem 13.  [3, Theorem 5.1].  Let  A  be a torsion-free abelian group of finite rank, and let  B  and  C  be arbitrary groups (not necessarily abelian).  If  $A \times B \cong A \times C$ , then there exists a positive integer  n  such that the direct product of  n  copies of  B  is isomorphic to the direct product of  n  copies of  C .

Proof.  Since  End(A)  is a torsion-free finite rank Z-algebra, it satisfies power-substitution by Theorem 10.  For the case where  B  and  C  are abelian, the result now follows from Corollary 4.  The general case is just a matter of observing that the appropriate version of Theorem 3 is valid in the category of all groups [3, Theorem 2.3].

The case  $A = Z$  of Theorem 13 was proved by Hirshon [5, Theorem 1], and a restricted version of this case was also proved by Warfield [9, Theorem 2.1].  In addition, Warfield proved the case of Theorem 13 in which  A,B,C  are all torsion-free abelian of finite rank (unpublished).

Theorem 14.  Let  B  and  C  be real vector bundles over a compact Hausdorff space  X .  If there exists a line bundle  A  such that  $A \oplus B \cong A \oplus C$ , then  $B \oplus B \cong C \oplus C$ .

Proof.  In view of the category equivalence between vector bundles over  X  and finitely generated projective C(X)-modules developed in [7], this is equivalent to the following problem:  If  A , B  and  C  are finitely generated projective C(X)-modules with  rank(A) = 1  and

$A \oplus B \cong A \oplus C$ , prove that $B^2 \cong C^2$ .

Since  A  is a  rank  1  projective,  $\text{End}(A) \cong C(X)$ , hence Corollary 8 shows that  $\text{End}(A)$  satisfies power-substitution.  Now $B^n \cong C^n$  for some  $n > 0$ , by Corollary 4.  Keeping track of the powers which occur in the proofs of Corollary 8 and Theorem 3 reveals that actually $n = 2$ .

We do not know whether a higher-dimensional version of Theorem 14 holds, i.e., if  A  is a vector bundle of dimension  k , then is $B^n \cong C^n$  for some integer  n  (depending only on  k )?  To prove such a result would require proving a bounded form of power-substitution for matrix rings over  $C(X)$ , which is an open problem.

§5.  **Problems**.

(A)  Is the power-substitution property left-right symmetric?

(B)  Is it Morita-invariant?

(C)  Does Theorem 13 hold for finite rank abelian groups which are not torsion-free?  In particular, does the endomorphism ring of such a group satisfy power-substitution?  (The answer to both questions is yes in case the torsion subgroup of the group is a direct summand.)

(D)  Does a noncommutative algebraic Q-algebra satisfy power-substitution?  ( Q  denotes the rationals.)

(E)  Presumably Theorem 10 can be generalized to direct limits of finite rank algebras over some domains other than  Z .  Perhaps it would

work for a Dedekind domain  S  such that for all nonzero  b  in  S ,
the group of units of  S/bS  is torsion.

(F)  Does  $M_n(C(X))$  satisfy power-substitution?

# REFERENCES

1.  Evans, E. G., Jr.  "Krull-Schmidt and cancellation over local rings," _Pacific J. Math._ 46(1973), 115-121.

2.  Fuchs, L. "On a substitution property for modules," _Monatshefte für Math._ 75(1971), 198-204.

3.  Goodearl, K. R.  "Power-cancellation of groups and modules," _Pacific J. Math._, to appear.

4.  Goodearl, K. R. and R. B. Warfield, Jr.  "Algebras over zero-dimensional rings," to appear.

5.  Hirshon, R.  "The cancellation of an infinite cyclic group in direct products," _Arch. der Math._ 26(1975), 134-138.

6.  Kaplansky, I.  _Infinite Abelian Groups_ (revised edition), Univ. of Michigan Press, Ann Arbor, 1969.

7.  Swan, R. G.  "Vector bundles and projective modules," _Trans. Amer. Math. Soc._ 105(1962), 264-277.

8.  _____.  Algebraic K-Theory_, Math. Lecture Notes #76, Springer-Verlag, Berlin, 1968.

9.  Warfield, R. B., Jr. "Genus and cancellation for groups with finite commutator subgroup," _J. Pure and Applied Algebra_ 6(1975), 125-132.

10. _____.  "Notes on cancellation, stable range, and related topics," Univ. of Washington, August 1975.

DIAGRAMMATIC TECHNIQUES IN THE

STUDY OF INDECOMPOSABLE MODULES

Edward L. Green

University of Illinois

§1.  **Introduction**.  The first five sections of this paper summarize

recent "diagrammatic" results in the study of the module categories of

Artin rings.  Recall that an additive category  $A$  is of **finite** **repre-**

**sentation** **type** if there are only a finite number of nonisomorphic in-

decomposable objects in  $A$ .  A ring  R  is said to be of **finite** **repre-**

**sentation** **type** if  mod(R) , the category of finitely presented left R-

modules is of finite representation type.  All rings have a unit and all

modules, unless otherwise stated, are left modules.  An **Artin** **algebra**

R  is an Artin ring which is a finitely generated module over its center.

Thus, k-algebras which are finite dimensional over  k  are Artin algebras.

Loosely speaking, the results can be described as follows.  To each

radical squared zero Artin k-algebra and to each ring in a class of

hereditary k-algebras, we associate a finite directed graph.  This graph

and its representations contain all the information about the ring and

its module category (in that, there is a stable equivalence of categories

between the module category and the category of representations of the

graph).

149

The last four sections announce some results that will appear in
[14]. Except for Section 10, which is directly connected with the first
four sections, there is no immediate connection between Sections 7-9 and
2-4. In fact, though, many of the results in [14] are shown to be the
"best possible" by examples which employ the results and techniques
described in the beginning sections. Furthermore, many of the results
were found by detailed analysis of the structure of indecomposable
modules using "diagrammatic techniques."

§2. **k-species and Quivers**. Let $k$ be a fixed field. By a **k-species**
$S = (K_i, {}_iM_j)_{i,j \in I}$ we mean a finite set $I$ , such that for each $i,j$ in $I$
$K_i$ is a division ring which is finite dimensional and central over $k$ ,
and ${}_iM_j$ is a $K_i$-$K_j$-bimodule which is finite dimensional over $k$ .
The notion of a "k-species" was introduced by Gabriel [11]. There
have been generalizations of k-species, most notably "modulations of
valued graphs" [8] but for simplicity, we will restrict our attention to
k-species.

Let $S$ be a fixed k-species. To $S$ we associate a finite
directed graph, $Q(S)$ , called the **quiver of** $S$ . Let $n_{ij} =$
$\dim_{K_i}({}_iM_j) \times \dim({}_iM_j)_{K_j}$ . Then $Q(S)$ has $I$ as its vertex set and
if $i,j \in I$ then there are $n_{ij}$-arrows from $j$ to $i$ . It should be
noted that this definition is slightly different than that found in [9].

A **representation of** $S$ is a tuple $(V_i, f_{ij})_{i,j \in I}$ such that for
each $i,j \in I$ , $V_i$ is a finite dimensional $K_i$-module and

$f_{ij} : {}_iM_j \otimes_{K_j} V_j \to V_i$ is a $K_i$-module morphism. The category of representations of the k-species $S$, $M(S)$, has as objects the representations of $S$ and as morphisms maps $(a_i) : (V_i, f_{ij}) \to (V'_i, f'_{ij})$, where for each $i$ in $I$, $a_i$ in $\text{Hom}_{K_i}(V_i, V'_i)$ the following diagram commutes:

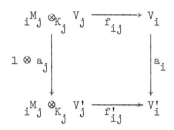

It is not hard to see that the category $M(S)$ is an abelian category with the obvious notion of direct sum. We will see that $M(S)$ is, in fact, a module category for a tensor algebra which is naturally defined from $S$.

§3. **The Classification Theorems.** The next result is the first classification theorem.

**Theorem A$_1$.** Let $S$ be a k-species. The category $M(S)$ is of finite representation type if and only if $Q(S)$ is composed of disjoint copies of the Dynkin diagrams:

$A_n$    $\overset{}{1} - \overset{}{2} - \cdots - \overset{}{n}$ ,   $n \geq 1$

$B_n$    $\overset{}{1} \mathrel{\underset{}{\overset{}{\Leftarrow}}} \overset{}{2} - \cdots - \overset{}{n}$ ,   $n \geq 2$

$C_n$    $\overset{}{1} \mathrel{\Rightarrow} \overset{}{2} - \cdots - \overset{}{n}$ ,   $n \geq 3$

$D_n$    $\overset{}{1} - \overset{\overset{\textstyle 3}{\mid}}{2} - \overset{}{4} - \cdots - \overset{}{n}$ ,   $n \geq 4$

$E_6$    $\overset{}{1} - \overset{}{2} - \overset{\overset{\textstyle 4}{\mid}}{3} - \overset{}{5} - \overset{}{6}$

$E_7$    $\overset{}{1} - \overset{}{2} - \overset{\overset{\textstyle 4}{\mid}}{3} - \overset{}{5} - \overset{}{6} - \overset{}{7}$

$E_8$    $\overset{}{1} - \overset{}{2} - \overset{\overset{\textstyle 4}{\mid}}{3} - \overset{}{5} - \overset{}{6} - \overset{}{7} - \overset{}{8}$

$F_4$    $\overset{}{1} - \overset{}{2} \mathrel{\Rightarrow} \overset{}{3} - \overset{}{4}$

$G_2$    $\overset{}{1} \mathrel{\Rrightarrow} \overset{}{2}$

where $\overset{}{i} - \overset{}{j}$ means either $\overset{}{i} \rightarrow \overset{}{j}$ or $\overset{}{i} \leftarrow \overset{}{j}$ and $\overset{\overset{\textstyle n \text{ lines}}{}}{\underset{i \quad j}{\bigodot}}$ means

$\underset{i \quad j}{\overset{\overset{\textstyle n \text{ arrows}}{}}{\rightrightarrows}}$ and $\dim_{K_i} {}_i M_j = 0$ . Furthermore, the number of nonisomorphic

indecomposable representations for these diagrams is given by the fol-

lowing table:

| Type | Number of Indecomposables |
|------|---------------------------|
| $A_n$ | $\frac{1}{2}n(n + 1)$ |
| $B_n$ | $n^2$ |
| $C_n$ | $n^2$ |
| $D_n$ | $n(n - 1)$ |
| $E_6$ | 36 |
| $E_7$ | 63 |
| $E_8$ | 120 |
| $F_4$ | 24 |
| $G_2$ | 6 |

Recently, representation type has been broken into three classes: finite, tame and wild. Before discussing the proof of Theorem $A_1$ we will state another classification theorem. A category $A$ is of **wild representation type** if there is a fully faithful functor $F : \text{mod}(k[X,Y]) \to A$, where $k[X,Y]$ is the polynomial ring over the field $k$ in two noncommuting indeterminates. It is easily shown [12] that $\text{mod}(k[X,Y])$ is "unclassifiable" in that a classification of the nonisomorphic indecomposables gives a classification of all finite dimensional A-modules, for each finitely generated k-algebra $A$. We say that a category $A$ is of **tame representation type** if it is not of wild representation type. The following theorem involves the "extended Dynkin diagrams," a list of which may be found in [8].

Theorem $A_2$.  Let  $S$  be a k-species.  Then  $M(S)$  is of tame representation type if and only if  $Q(S)$  is composed of disjoint copies of
Dynkin diagrams and extended Dynkin diagrams.

There have been basically three types of proofs given for Theorems
$A_1$ and $A_2$.  First there have been the computational and categorical
proofs [7,9,10].  These involve translating the problem to an appropriate category, for example, a category of filtered vector spaces, and
then computing all indecomposables in that category.  The second type
of proof involves Coxeter functors and quadratic forms [4,8,12].  In
this approach, the Weyl groups occur.  One also associates a quadratic
form to each k-species  $S$ .  If  $S = (K_i, {}_iM_j)_{i,j \in I}$  and for each  i   in
$I$ ,   $K_i = k$ , then the quadratic form is

$$q(S) = \sum_{i \in I} x_i^2 - \sum_{i,j \in I} n_{ij} x_i x_j .$$

Then  $S$  is of finite representation type if and only if  $q(S)$  is positive definite; and,  $S$  is of tame representation type if and only if
$q(S)$  is positive nondefinite.  Furthermore, the positive roots of
$q(S)$  correspond to the isomorphism classes of indecomposable representations of  $Q(S)$ .  Finally, a completely ring theoretic proof of the
classification of Artin algebras with radical square zero of finite
representation type was given by W. Muller [20].  I mention this here
since Muller obtains the diagrammatic results (see §5), without appealing
to representations of diagrams.

It is interesting to note that recent work of Auslander-Reiten
[2] and M. I. Platzeck [26] connects the last two approaches to "almost
split exact sequences" [3] and gives a unified and general proof classifying which Artin algebras, stably equivalent to hereditary Artin
algebras, are of finite representation type (see §4).


§4.  **k-species and Tensor Algebras.**  Let  $R = K_1 \times \cdots \times K_n$  be the
product ring of the  $K_i$  where each  $K_i$  is a division ring, finite
dimensional and central over  $k$  and where  $k$  acts centrally in the
product  $\Pi K_i$ .  Let  $M = \coprod_{i,j=1}^{n} {}_iM_j$ , where the  ${}_iM_j$  are $K_i$-$K_j$-bimodules
which are finite dimensional over  $k$ .  We view  $M$  as an R-R-bimodule.
Let  $T$  denote the tensor algebra  $R + M + M \otimes_R M + \otimes_R^3 M + \cdots$ .  Such
tensor algebras will be called  **special tensor algebras**.  To the special
tensor algebra  $T$  we may associate the k-species  $S(T) = (K_i, {}_iM_j)_{i,j \in I}$ ,
where  $I = \{1,\ldots,n\}$ .  Conversely, given a k-species  $S$  there is a
special tensor algebra  $T(S)$  associated to it.

For clarity, we give three examples.

(I)  Let  $I = \{1,\ldots,n\}$ ,  $K_i = k$  for all  $i$  and let

$${}_iM_j = \begin{cases} k, & \text{if } j = i - 1 \\ 0, & \text{otherwise} \end{cases}$$

. Then  $Q(S)$  is  $\overset{\bullet}{1} \to \overset{\bullet}{2} \to \cdots \to \overset{\bullet}{n}$ .

It is not hard to show that  $T(S)$  is isomorphic to the ring of  $n \times n$
lower triangular matrices over  $k$ , e.g.,

$$\begin{bmatrix} k & & & \\ k & k & 0 & \\ \vdots & & \ddots & \\ k & \cdots & & k \end{bmatrix} \quad .$$

(II) Let $I = \{1\}$ and $K_1 = k$ . Let $_1M_1 = V$ , where $V$ is $r$

copies of $k$ . Then $Q(S)$ is $\cdot$ $T(S) = k\{X_1,\ldots,X_r\}$ ,

the polynomial ring over $k$ in $r$ noncommuting variables.

(III) Let $I = \{1,2\}$ , $K_i = k$ for $i = 1,2$ and let

$$_iM_j = \begin{cases} K_2 \otimes_k V \otimes_k K_1 \ , & \text{if } i = 2 \text{ and } j = 1 \\ 0 & , \text{ otherwise} \end{cases} \quad , \text{ where } V \text{ is}$$

isomorphic to $r$ copies of $k$ . Then $Q(S)$ is .

It is not hard to show that $T(S)$ is isomorphic to the

generalized triangular matrix ring $\begin{bmatrix} k & 0 \\ V & k \end{bmatrix}$ .

Theorem B. Let $T$ be a special tensor algebra and let $S$ be the k-species associated to $T$ . Then $\text{mod}(T)$ is equivalent to $M(S)$ .

Proof. [7,15] We content ourselves with giving the equivalence
$F : \text{mod}(T) \to M(S)$ . Let $X$ be a T-module. Let $e_i = (0,\ldots,1,0,\ldots,0)$ ,
where 1 is in the i-th place, be an idempotent in $R$ . We view
$\{e_1,\ldots,e_n\}$ as a set of nonisomorphic idempotents in $T$ . Then
$F(X) = (V_i, f_{ij})$ where $V_i = e_iX$ , viewing $e_iX$ as a $K_i$-module and
$f_{ij} : {}_iM_j \otimes_{K_j} V_j \to V_i$ is induced from the multiplication map

$M \otimes_R X \to X$ . Finally, if $a : X \to X'$ is a T-morphism, define $F(a) =$ $(a_i)$ by $a_i$ is the restriction of $a$ to $e_i X \to e_i X'$ .

Theorems $A_1$ and B yield a classification of special tensor algebras of finite representation type. In fact, we get a classification of all hereditary k-algebras, finite dimensional over $k$ [7, Proposition 10.2]. As the sketch of the proof of Theorem B shows, there is an intimate connection between finitely generated T-modules and the representations of the associated k-species.

§5.  k-species and Radical Square Zero k-algebras.

Proposition C. [3] Let  A  be an Artin algebra with radical $\underline{a}$ . Suppose that $\underline{a}^2 = 0$ . Let  B  be the ring $\begin{pmatrix} A/\underline{a} & 0 \\ \underline{a} & A/\underline{a} \end{pmatrix}$ . Then  A  is of finite representation type if and only if  B  is of finite representation type.

The proof shows that there is a close connection between A-modules and B-modules. It is not hard to show that  B  is a hereditary Artin algebra. For the rest of this section let  A  be a k-algebra, finite dimensional over  k , with radical  $\underline{a}$  such that $\underline{a}^2 = 0$ . Let  B  be as in Proposition C.  Assume further that  A  is a basic ring, i.e.,

$A/\underline{a}$  is isomorphic to a product of division rings  $K_1 \times \cdots \times K_n$ , each

$K_i$  finite dimensional over  k .  One may easily show that in this case,

B  is isomorphic to a special tensor algebra.  There is a k-species

$S(B)$  associated to  B , which by Theorem $A_1$ classifies if  B , and

hence  A , is of finite representation type.  We define the <u>separated</u>

<u>quiver</u> <u>of</u>  A  [8,11], $Q'(A)$ , to be  $Q(S(B))$ .  We describe  $Q'(A)$

explicitly:


Let  $\{e_1,\ldots,e_n\}$  be a full set of nonisomorphic primitive idempotents

for  A .  Let  $K_i = e_i(A/\underline{a})e_i$ .  Let  $I = \{1,\ldots,n,1',\ldots,n'\}$ .  Let

$K_{i'} = K_i$ , and  $_\alpha M_\beta = \begin{cases} e_i \underline{a} e_j , & \text{if } \alpha = i' \text{ and } \beta = j \\ 0 , & \text{otherwise} \end{cases}$ .  Then

$S(B) = (K_\alpha, {}_\alpha M_\beta)_{\alpha,\beta \in I}$  and  $Q'(A)$  is the quiver associated  $S(B)$ .

   If  A  is not a basic ring, it is Morita equivalent to one, say

$A^*$ .  If we define  $Q'(A)$  to be  $Q'(A^*)$ , the above description of

$Q'(A)$  still holds.  Applying Theorem A, we get the next result.


<u>Theorem D</u>.  If  A  is a finite dimensional k-algebra with radical  $\underline{a}$

such that  $\underline{a}^2 = 0$ , then  A  is of finite representation type if and

only if  $Q'(A)$  is composed of disjoint copies of Dynkin diagrams.

§6. **Further Applications.** At this time, diagrammatic techniques have only led to the classification of rings of finite representation type in the hereditary and radical squared zero cases. Aside from the question of classifying rings of finite representation type, diagrammatic techniques have at least three other important applications. The first is in providing a tool useful in other classification problems [1,15,16, 17,19,27], and see Theorem 12 in Section 10. Secondly, since there is such a close connection between representations of diagrams and modules, one can often construct examples of rings having modules with desired properties [1,14,16,17,18]. Thirdly, the use of diagrammatic techniques gives a method of testing hypotheses or obtaining insight into the structure of indecomposable modules over Artin rings.

The interested reader is also referred to [5,6,23,25].

§7. **Theorems on Amalgamations.** The next four sections summarize a number of results that will appear in [14]. This section deals with the determination of criteria for amalgamations (pushouts) of indecomposable modules to be indecomposable. The first two results are valid in any module category.

**Theorem 1.** Let $M_1$ be indecomposable and let $a_i : A \to M_i$ be nonzero proper monomorphisms, $i = 1,2$, such that $\mathrm{Hom}(M_1, M_2/\mathrm{im}(a_2)) = 0$.

(a) If $M_2$ is indecomposable and $\mathrm{Hom}(M_2, M_1/\mathrm{im}(a_1)) = 0$ then the pushout $\text{coker}\begin{pmatrix} a_1 \\ a_2 \end{pmatrix}$ is indecomposable.

(b) If $M_2/\text{im}(a_2)$ is indecomposable then $\text{coker}\begin{pmatrix} a_1 \\ a_2 \end{pmatrix}$ is in-

decomposable if and only if there is no homomorphism $f : M_2 \rightarrow M_1$ such

that $fa_2 = a_1$ .

Let $R$ be a ring and let $B$ be an R-module of finite length. We say that $B$ is **basic** if $B/\text{rad } B$ is a simple R-module. We say that $B$ is **cobasic** if $\text{soc}(B)$ is an R-module.

**Theorem 2.** If $0 \longrightarrow S \longrightarrow B_1 \oplus B_2 \xrightarrow{(p_1, p_2)} M \longrightarrow$ is a nonsplit exact sequence where $S$ is a simple module and $B_1$ and $B_2$ are basic modules of finite length, then $M$ is indecomposable if and only if neither $p_1$ nor $p_2$ is a split monomorphism.

If $S$ is a simple submodule of a nonsimple basic module $B$ of finite length over a homomorphic image of a hereditary left Artin ring such that $\text{Ext}^1(B,B) = 0$ , then, using Theorem 2, it is readily checked that the cokernel of some homomorphism $S \rightarrow B \oplus B$ is in-decomposable if and only if $\text{Ext}^1(B/S,B) \neq 0$ . When $\text{Ext}^1(B,B) \neq 0$ , we have the next result.

**Theorem 3.** If $B$ is a basic module over a homomorphic image of a hereditary left Artin ring, then every nonsplit extension of $B$ by $B$ is indecomposable.

The final result of this section is useful in the proofs of
Theorems 11 and 12 following.  It enables one to construct with ease,
albeit in a restricted context, new indecomposable modules from known
ones.


<u>Theorem 4.</u>  Let $M_1, \ldots, M_{n+1}$ be indecomposable modules over a left
Artin hereditary ring of Loewy length 2 such that
$\mathrm{Hom}(M_i/\mathrm{rad}\ M_i, M_j/\mathrm{rad}\ M_j) = 0$ for $i < j$ .  Let $A_1, \ldots, A_n$ be nonzero
modules that admit superfluous monomorphisms $u_i : A_i \to M_i$ and
$v_i : A_i \to M_{i+1}$ .  Then coker(w) is indecomposable where

$w : \coprod\limits_1^n A_i \to \coprod\limits_1^{n+1} M_i$ is the matrix

$$
\begin{bmatrix}
u_1 & 0 & \cdots & & 0 \\
v_1 & u_2 & & & \\
0 & & \ddots & & 0 \\
\vdots & & & \ddots & \\
& & & v_{n-1} & u_n \\
0 & \cdots & 0 & & v_n
\end{bmatrix} .
$$

§8.  **Construction of Indecomposable Modules in Artin Rings.**  Let  R  be
a left Artin ring with radical  $\underline{r}$ .  Let  M  be an R-module.  The **core
of**  M ,  C(M) , is the intersection of the nonsuperfluous submodules of
M ; that is, it is the intersection of submodules  N  such that
N + N' = M  for some proper submodule  N'  of  M .  We define the **cocore
of**  M ,  $C^o(M)$ , dually; namely, the sum of the nonessential submodules
of  M .  Clearly, any module with a core (meaning  $C(M) \neq 0$ ) or cocore
( $C^o(M) \neq M$ )  is indecomposable.

**Theorem 5.** Let  R  be a left Artin ring such that some nonbasic R-module
has a core with nonsimple socle.  Then there exists a finitely generated
indecomposable R-module that does not have a cocore.

We note that the proof is a constructive proof.  In the following
result, we impose much more restrictive hypothesis than in Theorem 2.
This hypothesis occurs often and thus the result gives an interesting
method of constructing modules with cores.

**Theorem 6.**  Let  R  be a left Artin ring.  Let  $M = \text{coker} \begin{pmatrix} a_1 \\ a_2 \end{pmatrix}$  where
$a_i : S \to P_i$  are proper monomorphisms,  $i = 1,2$ ,  S  a simple R-module
and  $P_1$  and  $P_2$  are basic projective R-modules.  If  $P_1$  and  $P_2$  each
have at most one simple submodule isomorphic to  S  then either

    (a)  $C(M) \cong S$ ,

    (b)  there exists  $f : P_1 \to P_2$  such that  $fa_1 = a_2$ , or

(c)  there exists  $f : P_2 \to P_1$  such that  $fa_2 = a_1$ .

The following result provides a method, other than amalgamations, of constructing modules with cores.

Theorem 7.  Let  X  be a nonsplit extension of a simple R-module  S  by a module  M  with a core.

(a)  If the core of  M  contains a simple module not isomorphic to  S , then  X  has a core and hence is indecomposable.

(b)  If  X  is not indecomposable, then  $C(M) \cong S$  and every decomposition of  X  has the form  $X = X_1 \oplus X_2$  where  $X_1$  and  $X_2$  have cores.

We end this section with a theorem, whose proof is constructive, and which shows that most Artin rings have finitely generated indecomposable modules with zero core.

Theorem 8.  If a left Artin ring has a basic module containing three nonisomorphic simple submodules, then it has an indecomposable module of finite length with neither a core nor a cocore.

§9.  Classification of Modules with Cores.  We begin with some general results about modules with cores.

Theorem 9.  Let every indecomposable left injective module over a left

Artin ring have finite length.   Then

    (a)   every left module with a core has finite length,

    (b)   moreover, there is a bound on the lengths of modules with

cores.


It is worth noting that the hypothesis of Theorem 9 is satisfied

by Artin algebras.   Using Theorem 5, we obtain a generalization of a

result found in [1].   Recall that a module   M   has a **waist** if there is

a proper nonzero submodule   N   of   M   such that every submodule of   M

is either contained in or contains   N .


Theorem 10.   Every nonsimple indecomposable module over a left Artin

ring has a waist if and only if every indecomposable module is either

basic or cobasic.


Let   S   be a simple module over a radical squared zero Artin

algebra   R , let   V   be a submodule of the injective envelope of   S

properly containing   S , and let   $\coprod_{i=1}^{n+1} f_i : \coprod_{i=1}^{n+1} P_i \to V$   be a projective

cover of   V   where the   $P_i$   are indecomposable projective modules.

Then there exists maps   $e_i : S \to P_i$   such that   $g_i e_i = 1_S$ , where

$g_i = f_i \big|_{\text{im } e_i}$ .   Denoting the cokernel of

$$\begin{bmatrix} e_1 & 0 & \cdots & & 0 \\ -e_2 & e_2 & & & \\ 0 & \ddots & \ddots & & 0 \\ \vdots & & \ddots & \ddots & \\ & & & -e_n & e_n \\ 0 & \cdots & 0 & & -e_{n+1} \end{bmatrix} : \coprod_1^n S \to \coprod_1^{n+1} P_i$$

by $M(V)$ , we have the next result.

Theorem 11.  If  $R$  is a radical squared zero Artin algebra of finite representation type, then

(a)  $C(M(V)) = \begin{cases} P_1 & \text{if } V \text{ is a basic module } (n = 0) \\ S & \text{if } V \text{ is not a basic module} \end{cases}$ ,

(b)  if  $X$  is a nonsimple $R$-module with a core, then there is a unique  $V$  admitting a surjection  $w : M(V) \to X$  such that $\ker(w) \cong \mathrm{rad}\, M(V)$  and  $w(C(M(V))) \cong C(X)$ .

We remark that [7,20] show that the construction of  $M(V)$  depends only on  $V$ . We also remark that the theorem classifies modules with cores over radical squared zero Artin algebras of finite representation type.

§10.   Another Classification Theorem.   We denote the length of a module
M  by  $\ell(M)$ , the projective cover by  $P(M)$ , and the injective envelope
by  $E(M)$ .


Theorem 12.   The following are equivalent properties of a radical square
zero Artin algebra  R .

(a)   Every indecomposable module of finite length has a core or a
cocore.

(b)   Either every indecomposable module has a core, or else every
indecomposable module has a cocore.

(c)   (i)   Every indecomposable projective module and every indecomposable injective module has length at most  3 , and

(ii)   either  $\ell(E(\text{rad } Q)) \geq 5$  for each indecomposable projective  Q  such that  $\ell(Q) = 3$  or  $\ell(P(F/\text{rad } F)) \geq 5$  for each
indecomposable injective  F  such that  $\ell(F) = 3$ .

(d)   The separated diagram of  R  is a finite disjoint union of
diagrams  $A_1, \ldots, A_5$ ,  $B_2$  and  $C_3$ .

## REFERENCES

1. Auslander, M., E. L. Green and I. Reiten. "Modules with waists,"
   Ill. J. of Math. 19(#3)(1975), 467-478.

2. Auslander, M. and I. Reiten. "The dual of the transpose and
   Coxeter functors," to appear.

3. _____. "Representation theory of Artin
   algebras, I-III," Communications in Alg. 1974-75.

4. Bernstein, I. N., I. M. Gelfand and V. A. Ponomarev. "Coxeter
   functors and a theorem of Gabriel," Uspechi Mat. Nauk. 28(1973),
   19-33.

5. Brenner, S. "Quivers with commutativity conditions and some
   phenomenology of forms," Lecture Notes in Math. #488, Springer-
   Verlag, 1975.

6. Butler, M. C. R. "On the classification of local representations
   of finite abelian p-groups," Lecture Notes in Math. #488, Springer-
   Verlag, 1975.

7. Dlab, V. and C. Ringel. "On algebras of finite representation
   type," Carleton Math. Lecture Notes 2, 1973.

8. _____. "Representations of graphs and algebras,"
   Carleton Math. Lecture Notes 8, 1974.

9. Donovan, P. and M. Freislich. "The representation theory of finite
   graphs and associative algebras," Carleton Math. Lecture Notes 5,
   1973.

10. Gabriel, P.  "Unzerlegbare Darstellungen," Man. Math. 6(1972),
    71-103.

11. _____.  "Indecomposable representations II," Symposia Math.
    Ist. Naz. di Alt. Mat., Vol. XI, 1973.

12. _____.  "Représentations indécomposables," Séminaire Bourbaki
    #444(73/74).

13. Gelfand, I. M. and V. A. Ponomarev.  "Problems of linear algebra
    and classification of quadruples of subspaces in a finite dimen-
    sional vector space," Coll. Math. Soc. Bolyai, 5, Tihany, Hungary,
    1970.

14. Gordon, R. and E. I. Green.  "Modules with cores and amalgamations
    of indecomposable modules," to appear.

15. Green, E. L.  "The representation theory of tensor algebras," J.
    of Alg. 34(#1)(1975), 136-171.

16. _____.  "Smoothness and rigidity of tensor algebras and
    their factors," J. of Alg. 37(#3)(1975), 472-488.

17. _____.  "Complete intersections in tensor algebras I and II,"
    J. of Pure and Applied Alg. 7(1976), 317-332; 8(1976), 51-61.

18. Green, E. L. and W. Gustafson.  "Pathological quasi-Frobenius
    algebras of finite type," Comm. in Alg. 2(#3)(1974), 233-260.

19. Loupias, M.  "Indecomposable representations of finite ordered sets,"
    Lecture Notes in Math. #488, Springer-Verlag, 1975.

20. Muller, W.  "Unzerlegbare moduln uber artinschen ringen," Math.
    Z. 137(1974), 197-226.

21.  Nazarova, L. A.  "Representations of quadruples," <u>Isv</u>. <u>Akad</u>. <u>Nauk</u>. <u>SSSR</u>,
     1964, 1361-1378.

22.  _____.  "Representations of quivers of infinite type,"
     <u>Isv</u>. <u>Akad</u>. <u>Nauk</u>. <u>SSSR</u> 37(1973), 752-791.

23.  _____.  "Partially ordered sets with an infinite number
     of indecomposable representations," Lecture Notes in Math. #488,
     Springer-Verlag, 1975.

24.  Nazarova, L. A. and A. V. Roiter.  "Representations of ordered sets,"
     <u>Zapiski</u> <u>Nau</u>. <u>Sem</u>. <u>Leningr</u>. <u>Otd</u>. <u>Mat</u>. <u>Inst</u>. <u>Steklova</u> 28(1972),
     5-31.

25.  _____.  "Categorical matricial problems
     and the conjecture of Brauer-Thrall," preprint, Inst. of Math. of
     the Academy of SC. of Ukraine, Kiev, 1974.

26.  Platzeck, M. I.  "Representation theory of algebras stably equivalent
     to a hereditary Artin algebra," Thesis, Brandeis Univ., January 1976.

27.  Ringel, C.  "The representation type of local algebras," Lecture
     Notes in Math. #488, Springer-Verlag, 1975.

COHEN-MACAULAY RINGS, COMBINATORICS,

AND SIMPLICIAL COMPLEXES

Melvin Hochster[1]

Purdue University

University of Michigan

§1. __Introduction__. Let $\Delta$ be an (abstract) finite __simplicial__ __complex__ with vertices $x_0, \ldots, x_n$ (i.e., a family of subsets of $\{x_0, \ldots, x_n\}$ such that if $\sigma$ is in $\Delta$ and $\tau$ is a subset of $\sigma$ then $\tau$ is in $\Delta$, and such that $\{x_0\}, \ldots, \{x_n\}$ are in $\Delta$). Let $K$ be a commutative ring, usually, a field (throughout, all rings are commutative, with identity). Let $S = K[x_0, \ldots, x_n]$ be the polynomial ring generated by $x_0, \ldots, x_n$: by abuse of notation, we are regarding the vertices $x_i$ as indeterminates over $K$. Let $I_\Delta$ be the ideal generated by the monomials $x_{i_0} \ldots x_{i_r}$, $i_0 < \cdots < i_r$, such that $\{x_{i_0}, \ldots, x_{i_r}\}$ is not in $\Delta$. Let $K[\Delta] = S/I_\Delta$.

Recently G. Reisner [23] has characterized the Cohen-Macaulayness of $K[\Delta]$ in terms of purely topological properties of $\Delta$ and certain

---

[1] During the preparation of this paper, the author was supported by Purdue University, the University of Michigan, and a grant from the National Science Foundation.

subcomplexes (the links) of $\Delta$ . Oddly enough, the proof depends on results of [13] which study the action of the Frobenius on local cohomolo. The results are valid even when char(K) = 0 , but the proof for that case depends on "reduction" to characteristic $p > 0$ .

R. Stanley has used Reisner's result to prove the upper bound conjecture for triangulations of spheres (bounding the numbers of faces of various dimensions in a triangulation of an n-sphere in terms of  n and the number of vertices:  see (§2)).

We give here an expository account of Reisner's results (§2) (as well as certain background material about Cohen-Macaulay rings, Gorenstein rings projective resolutions, local cohomology, etc. in (§3)), and we give a brief sketch of how Stanley uses the Cohen-Macaulayness of  K[$\Delta$]  (§4). Moreover, we present several other applications of the ideas of Reisner's thesis:

(a)   determining the ranks of the free modules in a minimal
       (graded) free resolution of  $I_\Delta$ , i.e., the "Betti numbers"
       of  $I_\Delta$  (§5).

(b)   showing that the projective dimension of an ideal generated
       by monomials in a permutable R-sequence does not depend only
       on the "form" of the monomials, but on the specific R-
       sequence as well (§2).

(c)   characterizing when  K[$\Delta$]  is Gorenstein - in particular, showin.
       that any triangulation of a sphere is Gorenstein (§6).

(d)   determining when  S/I  is C-M (C-M ≡ Cohen-Macaulay) for an
       arbitrary ideal  I  generated by monomials in  $x_0, \ldots, x_n$

(not necessarily square-free any more) (§7).

(e)  recovering certain topological results from these ideas:

e.g., we show that Serre-Grothendieck duality on $\text{Proj}(K[\Delta])$

reduces to Poincaré duality on the geometric realization

$|\Delta|$ of $\Delta$ , if $|\Delta|$ is a manifold (this depends on (c) and

on studying $\text{Pic}(\text{Proj}(K[\Delta]))$ ) (§6).

The results of (§5), while new, are obtained by a routine elaboration of Reisner's methods.

(§6) and (§7) are entirely new, however. We note particularly that if $\dim(\Delta) \geq 2$ , then $K[\Delta]$ is Gorenstein for all Gorenstein $K$ if and only if $Z[\Delta]$ is Cohen-Macaulay and the links of the $(\dim(\Delta - 2))$-simplices of $\Delta$ are either circles, or lines with at most 3 vertices. [See the remarks, "Added in Proof," at the end.]

§2.  **Reisner's Results: Cohen-Macaulayness of** $K[\Delta]$ . Let $\mathbb{R}$ be the real numbers and let $\Delta$ be a given abstract simplicial complex with vertices $x_0, \dots, x_n$ . We identify $x_0, \dots, x_n$ with the standard basis for $\mathbb{R}^{n+1}$ , and let

$$|\Delta| = \bigcup_{\sigma \text{ in } \Delta} \text{convex hull }(\sigma) \; ;$$

$|\Delta| \subset \mathbb{R}^{n+1}$ is a topological space. If $i_0 < \cdots < i_r$ , and $\sigma = \{x_{i_0}, \dots, x_{i_r}\}$ is in $\Delta$ , then $\sigma$ is called an r-face of $\Delta$ .

$\underline{\dim}(\Delta) = \max\{r : \Delta \text{ has an r-face}\}$ = the topological dimension of $|\Delta|$ .

If $K$ is a ring, let $C.(\Delta)$ denote the simplicial chain complex with coefficients in $K$ (we write $C.(\Delta;K)$ if we need to be specific about what $K$ is):

$$C_i(\Delta) = \text{the free } K\text{-module on the } i\text{-faces of } \Delta$$

(thus $C_i(\Delta) = 0$ if $i > \dim(\Delta)$ or if $i \le -2$, but $C_{-1}(\Delta) \cong K$, since $\Delta$ has the unique $(-1)$-face $\emptyset$ ), while for each r-face $\sigma = \{x_{i_0}, \ldots, x_{i_r}\}$, $i_0 < \cdots < i_r$,

$$d(\sigma) = \Sigma_{j=0}^r (-1)^j (\sigma \backslash \{x_{i_j}\}),$$

where $\sigma \backslash \{x_{i_j}\}$ is the $(r-1)$-face obtained by deleting $x_{i_j}$ from $\sigma$. There is a dual complex $C^{\cdot}(\Delta; K)$, where

$$C^{\cdot}(\Delta) = \text{Hom}_K(C.(\Delta), K).$$

Note that if $f$ is in $\text{Hom}_K(C_i(\Delta), K) = C^i(\Delta)$, and $\delta$ is the differentiation, we have that for each $\sigma = \{x_{i_0}, \ldots, x_{i_{r+1}}\}$ in $\Delta$, $i_0 < \cdots < i_{r+1}$,

$$(\delta f)(\sigma) = \Sigma_{j=0}^{r+1} (-1)^j f(\sigma \backslash \{x_{i_j}\}).$$

The homology of $C.(\Delta)$ (resp., cohomology of $C^{\cdot}(\Delta)$ ) is denoted $\tilde{H}.(\Delta)$ (or $\tilde{H}.(\Delta; K)$, etc.) (resp., $\tilde{H}^{\cdot}(\Delta)$ ) and is called the __reduced__ simplicial homology (cohomology) of $\Delta$. If we omit $C_{-1}$ (resp., $C^{-1}$), replacing it by $0$, we get the (__nonreduced__) __simplicial (co)homology__, denoted $H.(\Delta)$ ( $H^{\cdot}(\Delta)$ ). Note that $\tilde{H}_i = H_i$ for $i \ge 1$, and similarly for cohomology. $H_i$, $H^i$ vanish for $i < 0$ and $i > \dim(r)$. If $K$ is a field, $\tilde{H}_i$ and $\tilde{H}^i$ are dual over $K$, and similarly for $H_i$, $H^i$.

N.B. By our definition, $\tilde{H}^i(\Delta) = 0$ if $i < 0$ __with one exception__, which is important for us:

$$\text{if } \Delta = \{\emptyset\} \ , \quad \tilde{H}^i(\Delta) \quad \begin{aligned} &= 0 && \text{if} && i \neq 1 \\ &\cong K && \text{if} && i = -1 \end{aligned} \quad .$$

Note that $\dim_K(H^0(\Delta))$ is the number of connected components, that if $\dim(\Delta) \geq 0$, then $\dim_K(\tilde{H}^0(\Delta)) = \dim_K(H^0(\Delta) - 1)$, and so that $\Delta$ is connected if and only if $\dim_K(\tilde{H}^0(\Delta)) = 0$, i.e., $\tilde{H}^0(\Delta) = 0$.

We note that $H_\cdot(\Delta)$, $\tilde{H}_\cdot(\Delta)$ and $H^\cdot(\Delta)$, $\tilde{H}^\cdot(\Delta)$ depend only on $|\Delta|$ : in fact, only on the homotopy class of $|\Delta|$. We refer the reader to any text on algebraic topology, e.g., [14] or [26] for more details.

If $\Delta$ is a simplicial complex (this always means finite here) and $\sigma$ in $\Delta$ is a face, we write $L_\Delta(\sigma) = L(\sigma)$, the __link__ of $\sigma$, for

$$\{\tau \text{ in } \Delta : \tau \cap \sigma = \emptyset \text{ and } \tau \cup \sigma \text{ is in } \Delta\} .$$

If $\sigma = \{x\}$ is a vertex, we write $L(\{x\}) = L(x)$. If $\Delta_0 \subset \Delta$ is a subcomplex, let $|\Delta_0|$ equal

$$\bigcup_{\sigma \text{ in } \Delta_0} \text{convex hull } (\sigma)$$

within $|\Delta|$. We note that if $x$ is a vertex and $C = \bigcup_{x \text{ in } \sigma} |\sigma|$, the __closed star__ of $x$, and $U = \{p \text{ in } C : p \text{ has a positive } x \text{ coordinate}\}$, the __open star__ of $x$, then $|L(x)| = C - U$. Moreover $C$ is the cone over $L(x)$.

Examples.   (a)

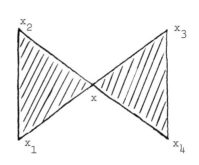

L(x) =

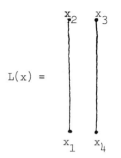

(b)

L(x) =

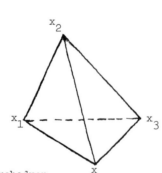

(c)

Δ = solid tetrahedron

$\Delta_i$ = i-skeleton,

   i = 2,1,0  .

$L_\Delta(x) =$

$L_{\Delta_2}(x) =$

$L_{\Delta_1}(x) =$

$L_{\Delta_0}(x) = \emptyset$

Note that if $\sigma = \{x_{i_0}, \ldots, x_{i_r}\}$, $i_0 < \cdots < i_r$, and $\sigma' = \sigma \backslash \{x_{i_r}\}$, then

$$L_\Delta(\sigma) = L_{L_\Delta(\sigma')}(x_{i_r}),$$

and iterating gives formulas like

$$L_\Delta(\{x_{i_0}, x_{i_1}, x_{i_2}\}) = L_{L_{L_\Delta(x_{i_0})}(x_{i_1})}(x_{i_2}).$$

Note that if $\sigma = \emptyset$, $L(\sigma) = \Delta$.

For general facts about Cohen-Macaulayness, we refer the reader to (§3). For the moment we recall simply that $R$ is C-M, or Cohen-Macaulay, if for every prime $P$, $\text{height}(P) = \text{depth}(P)$ (there is an $R$-sequence of length $\text{height}(P)$ in $P$).

Theorem (2.1). (G. Reisner). Let $K$ be a field, and $\Delta$ a finite simplicial complex. Then $K[\Delta]$ is C-M if and only if for every link $L$ of every $\sigma$ in $\Delta$, including $\Delta = L_\Delta(\emptyset)$,

$$\tilde{H}^i(L; K) = 0, \qquad 0 \leq i < \dim(L).$$

If $|\Delta|$ is a (closed) manifold, the condition on the links (other than $\Delta$) holds automatically. Hence,

Corollary (2.2). (G. Reisner). If $|\Delta|$ is a closed manifold, then $K[\Delta]$ is Cohen-Macaulay if and only if

$$\tilde{H}^i(\Delta;K) = 0 , \qquad 0 \le i < \dim(\Delta) .$$

To understand this result better, let us examine the connection between $K[\Delta]$ and $|\Delta|$ in greater detail.

We recall [4], [28] that the set of ideals $I$ generated by monomials in $x_0,\ldots,x_n$ is a distributive lattice under $+$, $\cap$, and each such $I$ is an irredundant intersection of ideals of the form

$\{x_{i_0}^{m_0},\ldots,x_{i_r}^{m_r}\}$ , $i_0 < \cdots < i_r$ . If $K$ is a domain, this decomposition is an irredundant primary decomposition. The ideals generated by the square-free monomials form a distributive sublattice, and if $K$ is reduced (no nilpotents but $0$ ), these are the radical ideals generated by monomials.

If $K$ is a field, $S = K[x_0,\ldots,x_n]$ , $I \subset S$ , let $V(I)$ be the subset of $K^{n+1}$ where the elements of $I$ vanish. If $I = \{x_{i_0},\ldots,x_{i_r}\}$ , we refer to $V(I)$ as a **coordinate hyperplane**. Let $\Sigma$ be the set of all subsets of $\{x_0,\ldots,x_n\}$ . Then we get one-one correspondences between:

(1) $S_1 = \{$subcomplexes of $\Sigma\}$ .

(2) $S_2 = \{$ideals generated by square-free monomials in $x_0,\ldots,x_n\}$ .

(3) $S_3 = \{$unions of coordinate hyperplanes$\}$ .

To go from $S_1$ to $S_2$ define $I_\Delta$ from $\Delta$ precisely as in the first paragraph of the Introduction. Henceforth, we relax the restriction

that each of $x_0,\ldots,x_n$ be a vertex of $\Delta$ : we allow some extra indeterminates corresponding to "extra" vertices (not occurring in $\Delta$ ).

We can go back from $S_2$ to $S_1$ by

$$I \mapsto \{\{x_{i_0},\ldots,x_{i_r}\} \subset \Sigma : x_{i_0}\ldots x_{i_r} \notin I\} .$$

The correspondence of $S_2$ to $S_3$ is $I \mapsto V(I)$ . We write
$H_\Delta$ for $V(I_\Delta)$ . Note that we can recover $\Delta$ from $H$ in $S_3$ , i.e.,
map $S_3$ to $S_1$ ) thus: $\{x_{i_0},\ldots,x_{i_r}\}$ is in $\Delta$ if and only if the
element of $K^{n+1}$ whose $x_{i_0},\ldots,x_{i_r}$ coordinates are $1$ and whose
other coordinates are $0$ is in $H$ .

If $K = \mathbb{R}$ , we can view the correspondence $S_3$ to $S_1$ as

$$H \mapsto H \cap |\Sigma| = |\Delta| .$$

Now assume that $x_0,\ldots,x_n$ occur as vertices of $\Delta$ . Then:

maximal faces of $\Delta$ correspond to minimal primes of $\Delta$

and

$$dim(K[\Delta]) = dim(H_\Delta)$$
$$= dim(\Delta + 1) .$$

Notice that if $K[\Delta]$ is C-M then $K[\Delta]$ has pure dimension
(all minimal primes have the same coheight), whence $\Delta$ is of pure
dimension (all maximal faces have the same dimension).

Let us say that $\Delta$ is C-M (over $K$ ) if $K[\Delta]$ is C-M.

The following are examples of the applications of Reisner's theorem.

0.  $\dim(\Delta) = 0$ .  $K[\Delta]$  is always C-M.

1.  $\dim(\Delta) = 1$ .  $K[\Delta]$  is C-M  if and only if  $\Delta$  is connected.

2.  $\dim(\Delta) = 2$ .  $K[\Delta]$  is C-M if and only if

    (a)  $\Delta$  has pure dimension.

    (b)  $\tilde{H}^i(\Delta) = 0$ ,  $i = 0,1$ .

    (c)  Every point of  $|\Delta|$  has arbitrarily small neighborhoods which are connected even with an arbitrary finite subset removed.

Note:  (c) follows from the condition that links be connected.

More specific examples:

A.   is <u>not</u> C-M ((c) fails at  x ).

Note:  $L(x) =$   is not C-M, for  $\tilde{H}^0(L(x)) \neq 0$ .

B.  2-spheres are C-M.

C.  A (hollow) torus   is <u>not</u> C-M, since  $\tilde{H}^1 \neq 0$ .

D.  A cylinder   is not C-M!  ( $\tilde{H}^1 \neq 0$ .)  (Of course, a circle is C-M.)

E.  A real projective plane is C-M if  $\mathrm{char}(K) \neq 2$  (for then  $\tilde{H}^1 = 0$ ).  But if  $\mathrm{char}(K) = 2$ ,  $\tilde{H}^1 \cong K$ , and the real projective plane is <u>not</u> C-M over  K .

If one looks at a few simple examples, one is tempted to believe that the projective resolution of an ideal generated by monomials in a permutable R-sequence depends only on the "form" of the set of monomials:

not on the R-sequence.  (See [4], [17], and [28] for some positive

results along these lines.)

However, as observed in [23], not only the form of the resolution

but even the projective dimension may vary, even for square-free monomials

in indeterminates over a field.  There is a dependence on the characteristic.

One gets examples from Example 2.E (the real projective plane) above:

take the ideal corresponding to any triangulation of the real projective

plane.

We make a few remarks about what happens when  K  is not a field.

We first note a trivial fact:  the Cohen-Macaulayness of  $K[\Delta]$ ,  K  a

field, depends only on  char(K) .  Let us write  $Z_p = Z/pZ$  when  p  is

a positive prime and  $Z_0 = Q$ , the rationals, i.e.,  $Z_p$  is the smallest

field of characteristic  p .


**Proposition (2.3).**  Let  K  be a commutative ring and  $\Delta$  a simplicial

complex.  The following conditions are equivalent:

    (a)  $K[\Delta]$  is C-M.

    (b)  K  is C-M and for every field  F  which is a K-algebra,

        $F[\Delta]$  is C-M.

    (c)  K  is C-M and  $Z_p[\Delta]$  is C-M for each positive prime integer

        p  such that  $1/p \notin K$ , and, if  $Z \subset K$ , for  p = 0  as well.


This follows easily from the following proposition.


**Proposition (2.4).**  (a)  The following conditions on a simplicial complex

Δ   are equivalent:

(1)   $Z[\Delta]$   is C-M.

(2)   $K[\Delta]$   is C-M for every C-M ring   K .

(3)   $Z_p[\Delta]$   is C-M for all primes   $p > 0$ .

(b)   Moreover, the following conditions are equivalent:

(1)   $Q[\Delta]$   is C-M.

(2)   $K[\Delta]$   is C-M for some field   K .

(3)   For all but finitely many positive prime integers

$p_1, \ldots, p_r$ ,   $K[\Delta]$   is C-M if   K   is a field with

char(K)   not in   $\{p_1, \ldots, p_r\}$ .


Part (b) can be deduced from "universal coefficient" theorems.

The case of prime fields is evidently crucial.  Or, as in algebraic

topology, one may use universal coefficient theorems to get the informa-

tion from the case   $K = Z$ .

For our purposes here it is simpler to stick with fields,

particularly positive characteristic fields - the Frobenius comes in

handy.


§3.  Depth, C-M Rings, Projective Resolutions, Local Cohomology, and

Gorenstein Rings.  This section is a lumping together of some basic

facts.  Let   R   be a Noetherian ring,   $I \subset R$   an ideal, and   M   an R-

module of finite type.  A sequence   $x_1, \ldots, x_n$   in   R   is called regular

on   M , an M-sequence, or an R-sequence on   M , if   $(x_1, \ldots, x_n)M \neq M$

and $x_{i+1}$ is not a zero divisor on $M/(x_1,\ldots,x_i)M$ , $0 \leq i < n$ .
If $IM \neq M$ , then $\text{depth}_I(M)$ , the depth of M on I , denotes the
length of any maximal M-sequence in I . By a theorem, any M-
sequence in I can be enlarged to one of length $\text{depth}_I(M)$ .

Proposition (3.1). If M,N are modules of finite type over a
Noetherian ring R with $\text{Ann}_R(M) = I$ , $\text{Ann}_R(N) = J$ , and $I + J \neq R$
(equivalently, $IN \neq N$ or $M \otimes_R N \neq 0$ ), then

$$\text{Ext}_R^i(M,N) = 0 \quad \text{if} \quad i < \text{depth}_I(N)$$
$$\text{Ext}_R^i(M,N) \neq 0 \quad \text{if} \quad i = \text{depth}_I(N) .$$

Theorem (3.2). (Auslander-Buchsbaum). Let $M \neq 0$ be a module of
finite type over a local ring $(R,m)$ . Then if M has a finite projective resolution, $\text{pd}_R(M)$ , the length of any shortest such resolution, is

$$\text{depth}_m(R) - \text{depth}_m(M) .$$

Proposition (3.3). (Rees). For any R-module $M \neq 0$ of finite type,
if $I = \text{Ann}_R(M)$ then

$$\text{depth}_I(R) \leq \text{pd}_R(M) .$$

Proof. Apply (3.1) with $N = R$ .

Let $(R,m)$ be either a local ring with residue class field $K = R/m$ or else let $R = R_0 \oplus R_1 \oplus R_2 \oplus \cdots$ be a finitely generated nonnegatively graded K-algebra, $K$ a field, such that $R_0 = K$ and $R_1$ generates $R$ over $K$, and let $m = R_1 \oplus R_2 \oplus \cdots$ . (We insert certain adjectives parenthetically in the sequel which are needed in the graded case but not in the local case.)

Let $M$ be a finitely generated (graded) R-module. Then $M$ has a _minimal_ (graded) resolution by finitely generated (graded) free modules (and maps of degree $0$ ), say

$$\cdots \longrightarrow F_i \xrightarrow{d_i} F_{i-1} \longrightarrow \cdots \longrightarrow F_1 \xrightarrow{d_1} F_0 \xrightarrow{d_0} M \longrightarrow 0 \; ,$$

where the minimality means, equivalently, that

(i) for each $i \geq 0$ , $d_i$ maps the free generators of $F_i$ onto a minimal (homogeneous) basis for $Im(d_i)$ ; or

(ii) for each $i \geq 1$ , $Im(d_i) \subset mF_{i-1}$ .

Regard $K$ as the R-module $R/m$ . The numbers $b_i = \dim_K(Tor_i^R(K,M))$ are called the **Betti numbers** of $M$ . Let $F. = \cdots \to F_i \to \cdots \to F_1 \to F_0 \to 0$ . Then

$$Tor_i^R(K,M) \cong H_i(K \otimes F.) = H_i(\cdots \xrightarrow{0} K \otimes F_i \xrightarrow{0} \cdots \longrightarrow K \otimes F_0 \xrightarrow{0} 0)$$

$$\cong K \otimes F_i \quad \text{(all the maps are } 0 \text{ by condition (ii) above),}$$

whence $b_i = \text{rank } F_i$ .

It is usually difficult, given a specific $R$ and $M$ to find an explicit minimal resolution of $M$ or even to determine the Betti numbers.

Let  $x_1,\ldots,x_n$  denote n elements of a ring  R .  Let  $R^n = Ru_1 \oplus \cdots \oplus Ru_n$  be the free module on  n  generators.  The exterior algebra  $\Lambda(R^n)$  becomes a free complex, called the __Koszul complex__  $K.(x_1,\ldots,x_n;R)$ ,

$$0 \longrightarrow \Lambda^n R^n \xrightarrow{d_n} \cdots \longrightarrow \Lambda^1(R^n) \xrightarrow{d_1} R \longrightarrow 0$$

where  $d_r(u_{i_0} \wedge \ldots u_{i_j} \wedge \ldots \wedge u_{i_r}) = \Sigma^r_{j=0} (-1)^j x_{i_j} (u_{i_0} \wedge \ldots \wedge \hat{u}_{i_j} \wedge \ldots \wedge u_{i_r})$

( $\hat{\ }$  denotes __omission__).  If  $\underline{x} = x_1,\ldots,x_n$  is an R-sequence, then  $K.(\underline{x};R)$  is acyclic: its augmentation is  $M = R/(x_1,\ldots,x_n)R$  and it gives a shortest projective resolution of  M .  If  R  is regular local and  $x_1,\ldots,x_n$  is a regular system of parameters, or if  $R = K[x_1,\ldots,x_n]$  and the  $x_i$  are indeterminates,  $K.(\underline{x};R)$  gives a minimal free resolution of  K , and hence may be used to compute  $Tor.^R(K, )$  and  $Ext_R^{\cdot}(K, )$ .

We briefly review what it means for a Noetherian ring  R  to be Cohen-Macaulay (C-M).

__Definition Proposition (3.4)__.  The following conditions on a Noetherian ring  R  are equivalent:

(1)  For every ideal  $I \subsetneq R$ ,  depth(I) = ht(I)  ( ht(I) = min{dim($R_P$) : $P \supset I$ , P prime} ).

(2)  For every maximal ideal  m ,  depth(m) = ht(m) .

(3)  For every maximal ideal  m , some system of parameters for  $R_m$  is an  $R_m$-sequence.

(4)  For every prime ideal  P , every system of parameters for  $R_P$  is an  $R_P$-sequence.

A ring  R  satisfying these equivalent conditons is called Cohen-Macaulay.

Theorem (3.5). Let  S  be a polynomial ring over a field  K  (resp., a regular local ring) and let  I  be a homogeneous (resp., arbitrary) proper ideal.  Then  $S/I$  is C-M if and only if  $\mathrm{pd}_S(S/I) = \mathrm{depth}(I)$ ( = height(I) ).

Moreover, in the graded case, if  m  is the homogeneous maximal ideal of  $R = S/I$ , then  R  is C-M if and only if  $R_m$  is C-M.

Proof.  The first part is essentially a consequence of (3.2) and the Cohen-Macaulayness of regular rings.  For the second part, see [19] or [11].

(3.6)  Examples of C-M and non-C-M rings.

0.  0-dimensional rings are C-M.

1.  A.  1-dimensional reduced rings are C-M.

B.  Some 1-dimensional nonreduced rings, e.g.,  $K[x,y]/(y^2)$ are C-M.

C.  $K[x,y]/(x^2,xy)$  is not C-M.

2.  A.  2-dimensional integrally closed rings are C-M, e.g., $K[x^2,xy,y^2] \subset K[x,y]$  is C-M.

B.  $K[x^2,x^3,y]$  is C-M, even though it is not integrally closed.

C.  $K[x^2,x^3,y,xy]$  is not C-M.

3.  Let  $R = K[x,y,z] = K[X,Y,Z]/(X^3 + Y^3 + Z^3)$ .  Then

K[xu,yu,zu,xv,yv,zv] $\subset$ R[u,v]  is integrally closed, 3-
dimensional, but not C-M.  See [7].

4.   A.   A regular ring is C-M.

     B.   If  R  is C-M and  $ht(x_1,\ldots,x_n) = n$ , then  $x_1,\ldots,x_n$
          is an R-sequence and  $R/(x_1,\ldots,x_n)$  is C-M.

     C.   If  R  is C-M,  $X = (x_{ij})$  is an  r  by  s  matrix of
          indeterminates over  R , and  I  is the ideal generated
          by the (t + 1)-size minors in  $S = R[x_{ij} : i,j]$ , then
          S/I  is C-M.  Moreover,  $ht(I) = (r - t)(s - t)$ .  See
          [10].

The notion of Cohen-Macaulayness is subtle and hard to understand.
It's an "arithmetic" condition with geometric consequences.

One key point is that ideals generated by R-sequences are unmixed.
Another is that when a Noetherian ring  R  can be represented as a finite
module over a regular ring  A ,  $A \subset R$ , then  R  is C-M if and only if
$R_m$  is $A_m$-free for each maximal ideal of  A .  Many theorems for "curves"
(1-dimensional reduced rings) can be generalized in a useful way to C-M
rings of arbitrary dimension.

One important example is <u>Serre-Grothendieck duality</u> which asserts:


<u>Theorem (3.7)</u>.  If  $(X,0_X)$  is a projective K-variety if pure dimension
r  and is C-M (all its local rings are C-M), then there is a coherent
sheaf  $\omega_X$  on  X , such that for every locally free coherent sheaf  F
on  X

$$H^i(X,F)^* \cong H^{r-i}(X,\underline{\omega}_X \otimes_{O_X} F^\vee) ,$$

where $^* = \text{Hom}_K(\ ,K)$ and $^\vee = \underline{\text{Hom}}_{O_X}(\ ,O_X)$ .

If $X$ is nonsingular $\underline{\omega}_X = \Omega^r_{X/K}$ = the sheaf of highest order Kähler differentials (the determinant of the cotangent bundle). If $X = \mathbb{P}^n$ , $\underline{\omega}_X = O_X(-n-1)$ . If $X$ is a closed subvariety of a non-singular n-dimensional $Y$ (e.g., $Y = \mathbb{P}^n$ ), $\underline{\omega}_X = \underline{\text{Ext}}^{n-r}_{O_Y}(O_X,\underline{\omega}_y)$ .


See [1] for further details.

This duality is a generalization of part of the Riemann-Roch theorem for curves.

For many other uses of C-M rings, e.g., the construction of "depth-sensitive" complexes analogous to the Koszul complex, vanishing of certain cohomology of sheaves, and extending sections of sheaves, we refer the reader to [10], [8], [9], [12], [13], and [25].

We next recall some basic facts about local cohomology. Let $R$ be Noetherian, $I$ an ideal, and $M$ an R-module. We define $H^{\cdot}_I(M) = \varinjlim_t \text{Ext}^{\cdot}_R(R/I^t,M)$ . If $x_1,\ldots,x_r$ generates an ideal with the same radical as $I$ , and $K^{\cdot}(\underline{x}^\infty;M)$ denotes the right complex

$$0 \to M \to \oplus M_{x_i} \to \oplus M_{x_i x_j} \to \cdots \to M_{x_i \ldots x_r} \to 0$$

$$= (\otimes_i (0 \to R \to R_{x_i} \to 0)) \otimes M ,$$

then $H^{\cdot}_I(M) \cong H^{\cdot}(K^{\cdot}(\underline{x}^\infty;M))$ . See [5], [12], and [13] for more details.

$H_I^{\cdot}(M)$ depends only on $\text{Rad}(I)$ . The following result (see [5])
summarizes some basic facts.

**Theorem (3.8).** Let $m$ be a maximal ideal of $R$ , and let $M$ be an R-
module of finite type. Let $I = \text{Ann}(M)$ , and assume $I \subseteq m$ (otherwise
$H_m^{\cdot}(M) = 0$ ). Then:

   (1)  The modules $H_m^{\cdot}(M)$ are Artinian (have D.C.C.). [Note: They
        need not have finite length!].

   (2)  Let $d = \text{depth}_m(M)$ and $h = \dim(R_m/I_m)$ . Then $H_m^d(M) \neq 0$ ,
        $H_m^h(M) \neq 0$ , but $H_m^i(M) = 0$ if $i < d$ or if $i > h$ .

   (3)  $H_m^{\cdot}(M) = H_{mR_m}^{\cdot}(M_m) = H_{m\hat{R}_m}^{\cdot}(\hat{M}_m)$ .

The fact that $H_m^{\cdot}(M)$ measures $\text{depth}_m(M)$ is particularly important.
In fact, quite generally:

$(3.8')$. If $M$ is Noetherian and $IM \neq M$ , the least integer $d$ such
that

$$H_I^d(M) \neq 0$$

is $\text{depth}_I(M)$ .

We conclude this section with a brief discussion of Gorenstein
rings. A Noetherian ring $R$ is called **Gorenstein** if $\text{id}_R(R)$ is
finite ( $\text{id}_R$ denotes the length of a shortest injective resolution).
The following characterization is often more useful.

Theorem (3.9). A Noetherian ring $R$ is Gorenstein if and only if $R$ is C-M and for every maximal (and even prime) ideal $m$ of $R$ and some (resp., every) system of parameters $x_1, \ldots, x_d$ for $R_m$, $R_m/(x_1, \ldots, x_d)R_m$ is injective as a module over itself.

We refer the reader to [2], [16] and [5] for details.

We note that a 0-dimensional local ring $(R,m)$ with $K = R/m$ is injective if and only if $\mathrm{Hom}_R(K,R) \cong K$. ( $\mathrm{Hom}_R(K,R) \cong \mathrm{Ann}_R(m)$ .) Moreover, if $R$ is C-M and $x_1, \ldots, x_d$ is a system of parameters,

$$\mathrm{Hom}_R(K, R/(x_1, \ldots, x_d)R) \cong \mathrm{Ext}_R^d(K,R) .$$

Thus we obtain the following.

Proposition (3.10). A local ring $R$ with residue class field $K$ is Gorenstein if and only if

$$\mathrm{Ext}_R^i(K,R) \cong \begin{cases} 0, & 0 \le i < \dim(R) \\ K, & i = \dim(R) \end{cases} .$$

In the graded case we may write $R = S/I$, where $S$ is a polynomial ring and $I$ is a homogeneous ideal. Likewise, if $R$ is local, in most cases which come up we may write $R = S/I$, where $S$ is regular local. In either case we obtain the following proposition.

Proposition (3.11). In either of the above cases, $R = S/I$ is Gorenstein

if and only if the highest nonzero Betti number $b_i$ for R occurs when
$i = ht(I)$ and is 1 .

<u>Proof</u>. The condition $b_j = 0$ for $j > ht(I) = depth_I(S) = d$ means
that $pd_S(R) = d$ , i.e., $depth_m(R) = dim(S - d) = dim(R)$ . Thus, this
part of the condition asserts that R is C-M. For the rest, we refer
the reader to [2].

It is worth noting that if R is C-M of pure dimension d ,
Spec(R) is connected, and R is a homomorphic image of a Gorenstein ring
S , one may associate with R the module $\Omega_R = Ext_S^{n-d}(R,S)$ , which is
unique up to tensoring with rank one projectives. If we restrict R to
being an algebra of finite type over a field K and S to being a
polynomial ring over K , or if we assume R is local, $\Omega_R$ is unique,
up to isomorphism.

We note that the restriction of $\underline{\omega}_X$ to an open affine Spec(R) of
X will correspond to a canonical module $\Omega_R$ for the coordinate ring of
that open affine.

One important fact we need for (§6) is that a C-M ring R is
Gorenstein if and only if $\Omega_R$ is a rank one projective.

In the situation of (3.7), a C-M X is (locally) Gorenstein if and
only if $\underline{\omega}_X$ is locally free of rank one.

§<u>4</u>.  <u>R. Stanley's Results on the Upper Bound Conjecture</u>.  The original

Upper Bound Conjecture (UBC), due to Motzkin [22], gives an upper bound

for the number of i-faces of a (d + 1)-dimensional convex polytope with

n + 1  vertices:  To wit, that the maximum possible is the number of

i-faces of a <u>cyclic</u> <u>polytope</u> , which is defined as the convex hull of

any  n + 1  distinct points on the curve

$$\{(t,t^2,\ldots,t^{d+1}) : t \text{ is in } \mathbb{R}\} .$$

This conjecture may be reduced to the simplicial case (the faces are

simplices) by a trick of "pulling vertices."  It then can be extended

to a conjecture about arbitrary triangulations of d-spheres or even

d-manifolds — and this is done, by Klee [17].

   In 1970 McMullen (see [20], [21]) proved UBC for polytopes using a

result of Brugesser and Mani [3] that the boundary complex of a convex

polytope is "shellable" in a certain strong sense.  However, not all

triangulations of spheres [30] are shellable (or even of cells [24]) and

the general question remained open, even for shperes.

   R. Stanley has solved UBC for spheres now and, more generally, for

all simplicial complexes  Δ  such that  K[Δ]  is C-M for some field  K

(equivalently, such that  Q[Δ]  is C-M).  See [27].

   We sketch the argument.  Let  $x_0,\ldots,x_n$  be indeterminates over

a field  K , let  $S = K[x_0,\ldots,x_n]$ , and let  I  be a homogeneous ideal

of  S .  Then  R = S/I  is a nonnegatively graded finitely generated

K-algebra generated by  $R_1$ , and with  $R_0 = K$ .  Such an  R  has a

Hilbert function  $H_R$  defined by

$$H_R(m) = \dim_K(R_m)$$

( $H_R$ agrees with a certain polynomial in $m$ of degree Krull

$\dim(R - 1)$ for all sufficiently large $m$ ).

When $R$ is C-M one can get inequalities on the values of $H_R$ by

computing in two different ways (here, Stanley is following Macaulay,

[18]). On the one hand, one can represent $R$ as a graded free module

over a polynomial subring generated by forms of equal degree, while on

the other hand one can use the fact that $R$ has a K-vector space basis

consisting of the images of certain monomials $x_1^{i_1} \ldots x_n^{i_n}$ such that if

the monomial $\nu$ occurs and $\mu | \nu$ (formally, in $S$ ), then $\mu$ occurs.

Now, when $R = K[\Delta] = S/I_\Delta$ it is very easy to compute $H_\Delta = H_{R_\Delta}$

([27], Proposition 3.2): in fact,

$$H_\Delta(m) = \begin{cases} 1 & \text{if} \quad m = 0 \\ \sum_{i=0}^{\dim(\Delta)} f_i \binom{m-1}{i} & \text{if} \quad m \geq 1 \,, \end{cases}$$

where $f_i$ is the number of i-faces of $\Delta$ .

The inequalities on the values of $H_\Delta$ for the C-M case mentioned

above yield UBC, in the form considered by McMullen, [20], [21], at oncce.

§5. The Betti Numbers of $K[\Delta]$ . Throughout this section $\Delta$ is a

simplicial complex with vertices $x_0, \ldots, x_n$ , unless otherwise specified

$K$ is a field, $S = K[x_0, \ldots, x_n]$ , and $R = K[\Delta] = S/I$ . Let $b_i$ be

the i-th Betti number of $R$ over $S$ , i.e., the rank of the i-th free module in a minimal graded free resolution of $R$ over $S$ . Thus,

$b_i = \dim_K(\text{Tor}_i^S(K,K[\Delta]))$ , where $K = S/(x_0,\ldots,x_n)$ .

If $T \subset \{x_0,\ldots,x_n\} = V$ , let $\Delta/T$ denote the largest subcomplex of $\Delta$ whose vertices are $V - T$ (i.e., $\Delta/T = \Delta \cap 2^{V-T}$ ).

Theorem (5.1). The i-th Betti number $b_i$ of $K[\Delta]$ over $S$ is given by

$$b_i = \sum_{\substack{-1 \leq q \leq n-i \\ T \subset V, \text{card}(T+q)=n-i}} \dim_K \tilde{H}^q(\Delta/T;K) \quad .$$

To see why this is true, we compute $\text{Tor}_{\cdot}^S(K,K[\Delta])$ by using the Koszul complex to resolve $K$ . Let $\Sigma$ be the simplicial complex consisting of all subsets of $V = \{x_0,\ldots,x_n\}$ , i.e., $\Sigma = 2^V$ . Then we may view $K_i = K_i(x_0,\ldots,x_n;S)$ as the free S-module on the $(n-i)$- faces of $\Sigma$ (where, if $V - \sigma = \{j_1,\ldots,j_i\}$ , $j_1 < \cdots < j_i$ , then $\sigma$ corresponds to $u_{j_1} \wedge \cdots \wedge u_{j_i}$ . In terms of this basis for $K_{\cdot}$ , we have that

$$d\sigma = \sum (-1)^{\alpha(j,\sigma)} x_j(\sigma \cup \{x_j\})$$

where the sum is taken over those $x_j$ in $V - \sigma$ , and where $\alpha(j,\sigma) = \text{card}\{t : t < j \text{ and } x_t \text{ in } V - \sigma\}$ .

If $\mu = x_{i_1}^{t_1} \cdots x_{i_h}^{t_h}$ in $S$ is a monomial, we write $\text{Supp}(\mu)$ for $\{x_i : x_i | \mu\} = \{x_{i_\nu} : t_\nu \geq 1\}$ . Then $\{\mu : \text{Supp}(\mu) \text{ is in } \Delta\}$ is a free K-basis for $K[\Delta]$ and $\{(\mu,\sigma) : \text{Supp}(\mu) \text{ is in } \Delta \text{ and } \sigma \text{ is an } (n-i)-$

face of $\Sigma$} is a free K-basis for $K[\Delta] \otimes_S K_i$ . Now,

$Tor_i^S(K,K[\Delta]) \cong H_i(K[\Delta] \otimes_S K.)$ and $d|_{K[\Delta] \otimes_S K_i}$ is given by

(*)                    $d(\mu,\sigma) = \Sigma(-1)^{\alpha(j,\sigma)}(x_j\mu,\sigma \cup \{x_j\})$ .

where the sum is taken over those $x_j$ in $V - \sigma$ with $\{x_j\} \cup Supp(\mu)$

in $\Delta$ .

We call $(\mu,\sigma)$ an **environment** if $Supp(\mu) \cap \sigma = \emptyset$ . Given $(\mu,\sigma)$ ,

we define an environment $env(\mu,\sigma)$ as $(\mu/\Pi x_i,\sigma - Supp(\mu))$ , where

the product is taken over those $x_i$ in $Supp(\mu) \cap \sigma$ . Note that if

$\{x_j\} \cup Supp(\mu)$ is in $\Delta$ , then

(#)                           $env(x_j\mu,\sigma \cup \{x_j\}) = env(\mu,\sigma)$ .

Following G. Reisner [23], we split the complex $K[\Delta] \otimes_S K.$ into

a direct sum of subcomplexes $L_.^\varepsilon$ , one for each environment $\varepsilon = (\nu,\tau)$ .

In fact, if $E$ is the set of environments and $\varepsilon$ is in $E$ , $L_0^\varepsilon$ is

spanned, as a free K-module, by the set

$$W_\varepsilon = \{(\mu,\sigma) : env(\mu,\sigma) = \varepsilon\} .$$

Clearly, $K[\Delta] \otimes_S K. = \oplus_{\varepsilon \text{ in } E} L_.^\varepsilon$ as K-modules. But d maps

$L_.^\varepsilon$ into itself, by (*) and (#), and we have a splitting as complexes.

Let $T_i^\varepsilon(\Delta) = H_i(L_.^\varepsilon)$ . Thus, $Tor_i^S(K,K[\Delta]) = \oplus_{\varepsilon \text{ in } E} T_i^\varepsilon(\Delta)$ , and it

remains to compute $T_.^\varepsilon(\Delta)$ .

We next note that each element $(\mu,\sigma)$ in $W_\varepsilon$ is uniquely determined,

given $E = (\nu,\tau)$ , by $\rho = \sigma - \tau$ . In fact, if we define

$$\rho_\varepsilon = (\nu\Pi_{x_j \text{ in } \rho} x_j, \tau \cup \rho) ,$$

then $(\mu,\sigma) = \rho_\varepsilon$ , where $\varepsilon = \text{env}(\mu,\sigma)$ and $\rho = \sigma - \tau$ . Let $\Delta_\varepsilon = \{\rho \in$

$\Delta : \text{Supp}(\nu) \cup \rho \in \Delta$ and $\tau \cap \rho = \emptyset\}$ . $\Delta_\varepsilon$ is a subcomplex of $\Delta$ . For

each $\rho$ , let $\gamma(\rho) = \Sigma_{x_j \in \rho} \alpha(j,\tau)$ . Thus, if $x_j \notin \rho$ , $\rho' = \rho \cup \{x_j\}$ ,

then

$$\gamma(\rho') = \gamma(\rho) + \alpha(j,\tau) .$$

Now, $\rho \mapsto \rho_\varepsilon$ is a bijection of $\Delta_\varepsilon$ with a free basis for $L_\cdot^\varepsilon$ (i-faces

of $\Delta_\varepsilon$ map to a free basis for $L_{n-i-\text{card}(\tau)}^\varepsilon$ ). Let $\{\rho^*\}$ be the dual

basis for $C^i(\Delta_\varepsilon)$ to the basis of i-faces for $C_i(\Delta_\varepsilon)$ (i.e., $\rho^*(\sigma) =$

$\delta_{\rho\sigma}$ ). Thus, $\phi : \rho_\varepsilon \mapsto (-1)^{\alpha(\rho)}\rho^*$ is an isomorphism of $L_\cdot^\varepsilon$ with

$C^\cdot(\Delta_\varepsilon)$ as K-modules which takes $L_{n-i-\text{card}(\tau)}^\varepsilon$ isomorphically onto

$C^i(\Delta_\varepsilon)$ . Moreover, for each $i$ the diagram

$$
\begin{array}{ccc}
C^i(\Delta_\varepsilon) & \xrightarrow{\quad\delta\quad} & C^{i+1}(\Delta_\varepsilon) \\
\phi\uparrow & & \phi\uparrow \\
L_{n-i-\text{card}(\tau)}^\varepsilon & \xrightarrow{\quad d\quad} & L_{n-i-1-\text{card}(\tau)}^\varepsilon
\end{array}
$$

commutes: $\delta(\phi(\rho_\varepsilon)) = \delta((-1)^{\gamma(\rho)}\rho^*)$

$$= (-1)^{\gamma(\rho)}\Sigma (-1)^{\beta(j,\rho)}(\rho \cup \{x_j\})^* ,$$

where $\beta(\rho,j) = \text{card}\{t < j : x_t \text{ in } \rho\} = \alpha(j,\tau) - \alpha(j,\tau \cup \rho)$ , i.e.,

(with all sums taken over $x_j$ not in $\rho$ , $\rho \cup \{x_j\}$ in $\Delta_\varepsilon$ )

$$\delta(\phi(\rho_\varepsilon)) = (-1)^{\gamma(\rho)}\Sigma_{x_j} (-1)^{\alpha(j,\tau)-\alpha(j,\tau\cup\rho)}(-1)^{\gamma(\rho\cup\{x_j\})}\phi((\rho \cup \{x_j\})_\varepsilon)$$

$$= \phi((-1)^{\gamma(\rho)}\Sigma_{x_j} (-1)^{\alpha(j,\tau)-\alpha(j,\tau\cup\rho)+\gamma(\rho)+\alpha(j,\tau)}(\rho \cup \{x_j\})_\varepsilon)$$

$$= \phi(\Sigma_{x_j} (-1)^{\alpha(j,\tau U\rho)} (\rho \cup \{x_j\})_\varepsilon) = \phi(d\rho_\varepsilon) \ , \text{ since}$$

$$d\rho_\varepsilon = d(\nu \Pi_{x_h \text{ in } \rho} x_h, \tau \cup \rho) = \text{(by (*))}$$

$$\Sigma_{(\{x_j\} \cup Supp(\nu \Pi_{x_h \text{ in } \rho} x_h)) \text{ in } \Delta} (-1)^{\alpha(j,\tau U\rho)} (x_j \nu \Pi_{x_h \text{ in } \rho} x_h, \tau \cup \rho \cup \{x_j\})$$
$$x_j \text{ not in } \tau \cup \rho$$

$$= \Sigma_{(\{x_j\} \cup Supp(\nu) \cup \rho) \text{ in } \Delta} (-1)^{\alpha(j,\tau U\rho)} (\rho \cup \{x_j\})_\varepsilon$$
$$x_j \text{ not in } \tau \cup \rho$$

$$= \Sigma_{x_j \text{ not in } \rho, \rho \cup \{x_j\}} \quad \text{ in } \Delta_\varepsilon \quad (-1)^{\alpha(j,\tau U\rho)} (\rho \cup \{x_j\})_\varepsilon \ , \text{ as}$$

required. Thus, $\phi$ induces isomorphisms

(**)                               $H_{n-i-card(\tau)}(L_\cdot^\varepsilon) \cong \tilde{H}^i(\Delta_\varepsilon)$ .

Now, if $\nu \neq 1$ , say $x_h$ is in $Supp(\nu)$ , then $\Delta_\varepsilon$ is a cone with $x_h$ as vertex: if $\rho$ is in $\Delta_\varepsilon$ , so does $\rho \cup \{x_h\}$ . In this case, $\tilde{H}^\cdot(\Delta_\varepsilon) = 0$ , and so $\Psi_\cdot^\varepsilon(\Delta) = H_\cdot(L_\cdot^c) = 0$ . If $\nu = 1$ , $\varepsilon = (1,\tau)$ , then $\Delta_\varepsilon = \Delta/\tau$ . Now, this entire discussion is valid whether $K$ is a field or not, so that we now have shown the following.

<u>Theorem (5.2)</u>. For any ring $K$ and any $\Delta$ ,

$$Tor_i^S(K, K[\Delta]) \cong \oplus_{\substack{-1 \leq q \leq n-i \\ T \subset V, card(T)+q=n-i}} \tilde{H}^q(\Delta/T;K)$$

and

$$\text{Ext}_S^{j+1}(K,K[\Delta]) \cong \bigoplus_{\substack{-1 \leq q \leq j \\ T \subset V, \text{card}(T)+q=j}} \tilde{H}^q(\Delta/T;K) \ .$$

The first statement is clear since $\text{Tor}_i^S(K,K[\Delta]) = \bigoplus_\varepsilon$ in $E$ $H_i(L_\cdot^\varepsilon) = \bigoplus_{(1,\tau)}$ in $E$ $\tilde{H}^{n-i-\text{card}(\tau)}(\Delta/\tau;K)$ : let $q = n - i-\text{card}(\tau)$ and $T = \tau$ . The second statement follows from the identification $\text{Ext}_S^{j+1}(K,N) \cong \text{Tor}_{n-j}^S(K,N)$ ( $N$ arbitrary). (For $\text{Ext}_S^{j+1}(K,N) = H^{j+1}(\text{Hom}_S(K_\cdot,N)) = H^{j+1}(\text{Hom}_S(K_\cdot,S) \otimes_S N)$ and $\text{Hom}_S(K_\cdot,S)$ , with the numbering reversed, is identical with $K_\cdot$ , so that

$$H^{j+1}(\text{Hom}_S(K_\cdot,S) \otimes_S N) \cong H_{n+1-(j+1)}(K_\cdot \otimes_S N) \cong \text{Tor}_{n-j}^S(K,N) \ .$$

<u>Corollary (5.3)</u>. Let $K$ be a field. Then

$$\text{depth}(K[\Delta]) \geq d$$

(<u>depth</u> means depth on the homogeneous maximal ideal) if and only if for all $q$ and $T \subset V$ with $-1 \leq q < d - 1$ and $q + \text{card}(T) < d - 1$

$$\tilde{H}^q(\Delta/T;K) = 0 \ .$$

Hence, $K[\Delta]$ is Cohen-Macaulay if and only if for all $q$ and $T \subset V$ with $-1 \leq q < \dim(\Delta)$ and $q + \text{card}(T) < \dim(\Delta)$ ,

$$\tilde{H}^q(\Delta/T;K) = 0 \ .$$

Proof. The first assertion is immediate from Theorem (5.2) and
Proposition (3.1). The second follows from the fact that  $\dim(K[\Delta]) =$
$\dim(\Delta) + 1$ .


The characterization of Cohen-Macaulayness given here is not nearly
as useful in practice as Reisner's link characterization.

We also have:


Proposition (5.4). Let  K  be a field. If  $K[\Delta]$  is Cohen-Macaulay and
$\dim(\Delta) = d + 1$ , then  $K[\Delta]$  is Gorenstein if and only if

$$\bigoplus_{\substack{-1 \leq q \leq d \\ T \subset V, \, \mathrm{card}(T) = d-q}} \tilde{H}^q(\Delta/T) \cong K \quad .$$


Proof. By Proposition (3.11),  $K[\Delta]$  will be Gorenstein if and only if
$b_{(n+1)-(d+1)} = b_{n-d} = 1$ , i.e., if and only if  $\mathrm{Tor}^S_{n-d}(K, K[\Delta]) \cong K$ .
The result is the immediate from Theorem (5.2).


Proposition (5.5). (a) Let  $\Delta$  be 0-dimensional and consist of  $n + 1$
points,  $n \geq 0$ . Then  $K[\Delta]$  is Gorenstein if and only if  $n \leq 1$ .

(b) Let  $\Delta$  be 1-dimensional, with  $(n + 1)$  vertices. Then  $\Delta$
is Gorenstein if and only if either  $|\Delta|$  is a circle or  $1 \leq n \leq 2$  and
$|\Delta|$  is a line.


Before proving this, we note the following.

Proposition (5.6). (a) Let x be a vertex of $\Delta$ . Then

$$K[\Delta]_x \cong K[L_\Delta(x)][X,1/X]$$

where X is an indeterminate.

(b) If $K[\Delta]$ is C-M (respectively, Gorenstein), so is $L_\Delta(\sigma)$ for every $\sigma$ in $\Delta$ .

Proof. (a) $K[\Delta]_x$ is graded, and every nonzero form can be written uniquely as $F(x/1)^t$ , where $\deg(F) = 0$ and $t = \deg(F)$ in $Z$ . Let R be the subring of degree 0 elements of $K[\Delta]_x$ . Then $K[\Delta]_x = R[X,1/X]$ where X is the image of x . R is spanned over K by the monomials $(\bar{x}_{i_1}/\bar{x})^{h_1}...(\bar{x}_{i_\nu}/\bar{x})^{h_\nu}$ , $x_{i_r} \neq x$ , where $^-$ denotes reduction modulo $I_\Delta$ , and the only relations on the $\bar{x}_i/\bar{x}$ are that $(\bar{x}_{i_1}/\bar{x})...(\bar{x}_{i_\nu}/\bar{x}) = 0$ if $\bar{x}_{1}...\bar{x}_{i_\nu}\bar{x} = 0$ in $K[\Delta]$ , i.e., if $\{x_{i_1},...,x_{i_\nu},x\}$ is not in $\Delta$ , or, equivalently, $\{x_{i_1},...,x_{i_\nu}\}$ is not in $L_\Delta(x)$ . Thus, $R = K[\bar{x}_i/\bar{x} : x_i \neq x] \cong K[L_\Delta(x)]$ .

(b) $K[\Delta]$ C-M or Gorenstein implies $K[\Delta]_x \cong K[L_\Delta(x)][X,1/X]$ is as well, and hence, so is $K[L_\Delta(x)][X,1/X]/(X - 1) = K[L_\Delta(x)]$ . This proves the result for links of vertices. The general result now follows, since each $L_\Delta(\sigma)$ is an iterated link of vertices.

Proof of (5.5). Let the vertices of $\Delta$ be $\{x_0,...,x_n\}$ .

(a) $d = 0$ . If $n = 0$ , $\tilde{H}^{-1}(\emptyset) + \tilde{H}^0(\Delta) \cong K \oplus 0 \cong K$ . If $n \geq 1$ , we have $\oplus_{x_i} \tilde{H}^{-1}(\Delta - \{x_i\}) \oplus \tilde{H}^0(\Delta) \cong (0^{n+1}) \oplus (K^n) \cong K^n$ . Thus, $K[\Delta]$

is Gorenstein if $n = 1$ but not if $n \geq 2$.

(b)  $d = 1$.  We have

$$[\oplus_{x_i \neq x_j} \tilde{H}^{-1}(\Delta/\{x_i,x_j\})] \oplus [\oplus_{x_i} \tilde{H}^0(\Delta/x_i)] \oplus [\tilde{H}^1(\Delta)] .$$

$K[\Delta]$  C-M if and only if  $\Delta$  is connected.  In order for  $K[\Delta]$  to be
Gorenstein, we must have  $L_\Delta(x_i)$  is Gorenstein for each  $i$ , i.e., at
most two edges meet at each vertex.  Thus,  $\Delta$  is not Gorenstein unless
it is connected and at most two edges meet at each vertex.  But this
says that  $|\Delta|$   is either a circle (and then  $n \geq 2$ ) or a line.

If  $|\Delta|$  is a circle then each  $\Delta/\{x_i,x_j\} \neq \emptyset$ , so that the first
bracketed term vanishes, each  $\Delta/\{x_i\}$  is a line, so that the second
bracketed term vanishes, while  $\tilde{H}^1(\Delta) \cong K$ .  Thus, circles are Gorenstein.

Now suppose  $|\Delta|$  is a line with endpoints, say,  $x_0, x_n$ .  If
$n = 1$  the sum becomes

$$[K] \oplus [0] \oplus [0] \cong K .$$

If  $n = 2$  the sum becomes

$$[0 \oplus 0 \oplus 0] \oplus [0 \oplus K \oplus 0] \oplus [0] \cong K .$$

(Note:  $\tilde{H}^0(\Delta/\{x_i\}) \cong K$ .)

If  $n \geq 3$  then  $n - 1 \neq 1$  and

$$\tilde{H}^0(\Delta/\{x_1\}) = \tilde{H}^0(\Delta/\{x_{n-1}\}) \cong K .$$

Thus, when $|\Delta|$ is a line, $K[\Delta]$ is Gorenstein precisely if there are at most $3$ vertices.

Remarks. Thus, while I conjecture that the Cohen-Macaulayness of $K[\Delta]$ is actually a topological property of $|\Delta|$, this is evidently false for the Gorenstein property. This situation will be clarified in (§6), where we study the Gorenstein property further.

§**6.** **Pic and Sheaf Cohomology on** $\text{Proj}(K[\Delta])$, **the Gorenstein Property for** $K[\Delta]$, **and Poincaré Duality on** $|\Delta|$. Let $K$ be a commutative ring, $\Delta$ a simplicial complex, and let $(X_{\Delta,K}, 0_{\Delta,K})$ or, briefly, $(X_\Delta, 0_\Delta)$, denote $\text{Proj}_K(K[\Delta])$.

We shall compute $\text{Pic}(X_\Delta)$ and then use the result to determine when $K[\Delta]$ is Gorenstein.

We say that $K$ is _seminormal_ if for all $n$ and indeterminates $u_1, \ldots, u_n$ over $K$,

$$\text{Pic}(K) \to \text{Pic}(K[u_1, \ldots, u_n])$$

is an isomorphism. (If $K$ is a ring, $\text{Pic}(K)$ denotes the group of isomorphism classes of rank one projectives under $\otimes$. More generally, if $X$ is scheme, $\text{Pic}(X)$ is the group of isomorphism classes of locally free sheaves of rank one under $\otimes$.)

Seminormality (defined differently) is studied in [29], where a number of equivalent properties are considered. If $K$ is reduced pseudogeometric Noetherian, then [6] K is seminormal if and only if

for each $\lambda$ in the total quotient ring of K , if $\lambda^2, \lambda^3$ are in K

then $\lambda$ is in K . In particular, if K is reduced pseudogeometric

Noetherian, then K normal implies K seminormal.


**Theorem (6.1).** Let K be a seminormal reduced ring with Pic(K) = 0 .
Then for every $\Delta$ , Pic(K[$\Delta$]) = 0 .


Before proving this we need the following.


**Lemma (6.2).** Let $I_1$ , $I_2$ be ideals of R , and suppose $U(R/I_1 \cap I_2) \to$
$U(R/I_1 + I_2)$ is surjective, where U(S) denotes the group of units of
the ring S . If $Pic(R/I_m) = 0$ , m = 1,2 , then $Pic(R/I_1 \cap I_2) = 0$ .


**Proof.** We may replace R by $R/I_1 \cap I_2$ , $I_m$ by $I_m/I_1 \cap I_2$ , and

hence assume $I_1 \cap I_2 = (0)$ . Let P be a rank one projective R-module.

$P \otimes R/I_m \cong P/I_m P$ is $(R/I_m)$-free and we may choose $p_m$ in P so

that its image in $P/I_m P$ is a free generator. Hence, the images of

$p_1, p_2$ in $P \otimes R/I_1 \otimes R/I_2 \cong P/(I_1 + I_2)P$ are both free generators, and

there is a unit $\alpha$ of $R/(I_1 + I_2)$ such that $p_1 \equiv \alpha p_2 \mod(I_1 + I_2)P$ .

Since the map of units if surjective, we have that $\alpha = \bar{a}$ for a unit

a of R , and then $p_1 - ap_2 = i_1 - i_2$ , where $i_m$ is in $I_m P$ .

Then $v = p_1 - i_1 = ap_2 - i_2$ has the property that its image in either

$P/I_1 P$ or $P/I_2 P$ is a free generator (we only need that a be invertible

modulo $I_2$ , in fact), so that $P = Rv + I_1 P = Rv + I_1(Rv + I_2 P) \subset Rv +$

$I_1 I_2 P = Rv$ , which implies that P is free.

<u>Proof of (6.1)</u>. Let $\Delta$ have vertices $x_0, \ldots, x_n$ , and let $S = K[x_0, \ldots, x_n]$ . We prove the result that $Pic(S/I) = 0$ for any ideal $I$ generated by square-free monomials in $x_0, \ldots, x_n$ , by induction on the size of the ideal: we assume the result for larger $I$ . Note that the result is trivial if $I = (x_0, \ldots, x_n)$ , and if $I$ is prime, then $I = (x_{i_0}, \ldots, x_{i_r})$ for some subset $\{x_{i_0}, \ldots, x_{i_r}\}$ of $\{x_0, \ldots, x_n\}$ , so that $S/I$ is a polynomial ring, and the result follows from the facts that $Pic(K) = 0$ and $K$ is seminormal.

Now suppose $I$ is not prime. Then $I = I_1 \cap I_2$ where $I_1, I_2$ are strictly larger ideals of the same form (e.g., $I_1$ can be taken as one prime from the primary decomposition and $I_2$ as the intersection of the others). Since $I_1, I_2$ are strictly larger, the theorem follows from Lemma (6.2) ( $Pic(S/I_m) = 0$ , $m = 1,2$ , by the induction hypothesis once we know that $U(S/I) \to U(S/I_1 + I_2)$ is surjective. Now $I_1 + I_2$ is also a radical ideal generated by monomials. But for any such ideal $J$ , $U(K) \to U(S/J)$ is an isomorphism. In fact we leave it to the reader to check that if $R$ is a nonnegatively graded <u>reduced</u> K-algebra with $R_0 = K$ , then $U(K) \to U(R)$ is an isomorphism.

<u>Theorem (6.3)</u>. Let $K$ be seminormal reduced with $Pic(K) = 0$ and suppose $|\Delta|$ cannot be disconnected by removing finitely many points and $\dim(\Delta) \geq 1$ . Let $U(K)$ be the group of units of $K$ . Then

$$Pic(Proj(K[\Delta])) \cong H^1(\Delta; U(K)) \oplus Z.$$

where the copy of $Z$ is generated by $0_\Delta(1)$ , the very ample sheaf induced by the homogeneous coordinate ring $K[\Delta]$ .

Proof. Let $f_i$ be the image of $x_i$ in $R = K[\Delta]$ , let $X = \text{Proj}(K[\Delta])$ ,
and let $X_i = X_{x_i} = \text{Spec}([K[\Delta]_{x_i}]_0)$ , where $[\ ]_0 = 0^{\text{th}}$ graded piece and,
by Proposition $(5.6)(a)$ and its proof, $[R_{x_i}]_0 = [R_{f_i}]_0 \cong K[L_\Delta(x_i)]$ ,
whence, $\text{Pic}([R_{f_i}]_0) = 0$ (Theorem $(6.1)$). Thus, if $F$ is a locally
free sheaf of rank one on $X$ , $F|X_i$ corresponds to a free $[R_{f_i}]_0$-
module, and it follows that if $U = \{X_0,\ldots,X_n\}$ (an open cover of $X$ )
and $W$ is the sheaf of germs of units (i.e., if $Y$ is an open affine
then $W(Y) = U(\Gamma(Y,O_\Delta))$ then

$$\text{Pic}(X) \cong \overset{\check{V}1}{H}{}^1(U;W) .$$

Note that $X_{i_1} \cap \cdots \cap X_{i_r} = \emptyset$ if and only if $f_{i_1} \ldots f_{i_r} = 0$ if and
only if $x_{i_1} \ldots x_{i_r}$ is in $I_\Delta$ if and only if $\{x_{i_1},\ldots,x_{i_r}\}$ is not in $\Delta$ .
If $T_\nu$ denotes the open star of the vertex $x_\nu$ in $|\Delta|$ , we also have
$X_{i_1} \cap \cdots \cap X_{i_r} = \emptyset$ if and only if $T_{i_1} \cap \cdots \cap T_{i_\nu} = \emptyset$ .
We next want to compute $U([R_{f_i f_j}]_0)$ for $i \neq j$ . There are two
cases. If $f_i f_j = 0$ (i.e., $X_i \cap X_j = \emptyset$ ), then $U([R_{f_i f_j}]_0) =$
$U([0]_0) = U(0)$ is trivial ( $0$ denotes the zero ring). If $f_i f_j \neq 0$
(i.e., $X_i \cap X_j \neq \emptyset$ ) then it is not hard to see that $U([R_{f_i f_j}]_0) \cong$
$U(K) \oplus Z$ , where the copy of $Z$ is generated by the image of $x_i/x_j$ or
$f_i/f_j$ , i.e.,

$$U([R_{f_i f_j}]_0) = U(K)\{(f_i/f_j)^m : m \text{ in } Z\}$$

if $f_i f_j \neq 0$ . Thus, a Čech 1-cocycle is determined by a pair of func-

tions $\lambda, \mu$ on the set of pairs $(i,j)$ such that $i \neq j$ and

$X_i \cap X_j \neq \emptyset$ , where

$$\lambda(i,j) \text{ is in } U(K)$$

$$\mu(i,j) \text{ is in } Z$$

and satisfying

(i)   $\lambda$ determines a Čech 1-cocycle on $\{X_0,\dots,X_n\}$ with coefficients in the constant sheaf $U(K)$ .

(ii)   $\mu(i,j) = \mu(j,i)$ (when defined) and if $i,j,k$ are distinct
and $X_i \cap X_j \cap X_k = \emptyset$ , then $\mu(i,j) = \mu(j,k) = \mu(k,i)$ .

(The cocycle is then obtained by sending $(i,j) \rightarrow \lambda(i,j)(f_i/f_j)^{\mu(i,j)}$

if $i \neq j$ and $X_i \cap X_j \neq \emptyset$ . (i), (ii) are precisely what we need to

guarantee that this is a cocycle.)

The pairs $\{i,j\}$ such that $i \neq j$ and $X_i \cap X_j \neq \emptyset$ correspond

precisely to the edges (1-simplices) $\{x_i,x_j\}$ in $\Delta$ . Thus, we may

view $\mu$ as a function on the edges, with the property that it is constant

on the boundary of each 2-simplex of $\Delta$ . This easily implies that $\mu$

is constant on all the edges of any simplex $\sigma$ in $\Delta$ . Consider the

equivalence relation on the maximal simplices generated by requiring $\sigma \sim \tau$

if $\sigma$ and $\tau$ have an edge in common. Clearly, if two edges lie in

equivalent simplices, $\mu$ has the same value on both. For each

equivalence class $C$ of maximal simplices, let $C^*$ be the set of points

of $|\Delta|$ lying in one of these simplices. Then clearly, for distinct

equivalence classes $C_1, C_2$, $C_1^* \cap C_2^*$ is finite. Hence, if there
is more than one $C_i^*$ of positive dimension, we can disconnect $|\Delta|$ by
removing finitely many points. But no $C^*$ can be zero-dimensional, for,
since $\dim(\Delta) \geq 1$ and $|\Delta| - \emptyset = |\Delta|$ is connected, each maximal face
of $\Delta$ is positive dimensional.

Thus, there is only one equivalence class, and $\mu$ must be constant,
say $\mu(i,j) = m$ for all $(i,j)$ such that $i \neq j$ and $X_i \cap X_j \neq \emptyset$.

It follows that there is a unique Čech 1-cocycle $\lambda$ with values in
the constant sheaf $U(K)$ and a unique integer $m$ such that for all $i,j$

$$\gamma(i,j) = \lambda(i,j)(f_i/f_j)^m .$$

On the other hand, any coboundary actually comes from a 0-cochain with
values in the constant sheaf $U(K)$, for $U([R_{f_i}]_0) = U(K[L_\Delta(x_i)]) = U(K)$.
Moreover, the cocycle

$$(i,j) \to (f_i/f_j)^m$$

corresponds to $0_X(m) = 0_X(1)^{\otimes m}$. Thus,

$$\mathrm{Pic}(X) \cong \check{H}^1(U;W) \cong \check{H}^1(U;U(K)) \oplus Z$$

where the copy of $Z$ corresponds to the powers of $0_\Delta(1)$.

Let $T = \{T_0, \ldots, T_n\}$ be the open cover of $|\Delta|$ by open stars of
vertices. Since $T_{i_1} \cap \cdots \cap T_{i_r} = \emptyset$ if and only if $X_{i_1} \cap \cdots \cap X_{i_r} = \emptyset$
we have

$$\check{H}^1(U;U(K)) \cong \check{H}^1(T;U(K)) \cong \check{H}^1(|\Delta|;U(K)) \cong H^1(|\Delta|;U(K))$$

(simplicial or singular) $\cong H^1(\Delta;U(K))$ (simplicial).

We are now ready to apply this result to the study of when $K[\Delta]$ is Gorenstein. The case $\dim(\Delta) \leq 1$ was handled in §5.

Theorem (6.4). Let $K$ be a seminormal reduced Gorenstein ring with $\mathrm{Pic}(K) = 0$ and $\Delta$ a simplicial complex with vertices $x_0, \ldots, x_n$. Let $S = K[x_0, \ldots, x_n]$ and let

$$\underline{\omega}_\Delta = (\mathrm{Ext}_S^{n-d}(K[\Delta],S))^{\sim},$$

where we assume $d = \dim(\Delta) \geq 1$ ( $\underline{\omega}_\Delta$ is the canonical sheaf on $(X_\Delta, 0_\Delta) = \mathrm{Proj}(K[\Delta])$, if $X_\Delta$ is C-M). Then $K[\Delta]$ is Gorenstein if and only if the following two conditions hold:

    (1) $K[\Delta]$ is C-M.

    (2) For some integer $m$, $\underline{\omega}_\Delta \cong 0_\Delta(m)$.

Before proving this, we state and prove the following important corollary.

Corollary (6.5). If $K$ is seminormal reduced Gorenstein with $\mathrm{Pic}(K) = 0$ $\Delta$ is a simplicial complex with $\dim(\Delta) \geq 2$, $\Delta$ is C-M over $K$ with Gorenstein links of vertices, and $H^1(\Delta;U(K)) = 0$, then $K[\Delta]$ is Gorenstein.

Proof. The hypothesis on the links implies that $X_\Delta$ is (locally)

Gorenstein, i.e., that $\underline{\omega}_\Delta$ corresponds to an element of $\mathrm{Pic}(X_\Delta) \cong$ $H^1(\Delta;U(K)) \oplus Z \cong 0 \oplus Z \cong Z$ , by Theorem (6.3). (Note that since $K[\Delta]$ is C-M, $\dim(\Delta) \geq 2$ , the link of each vertex is connected: this implies easily that $|\Delta|$ cannot be disconnected by removing finitely many points.) Since this copy of $Z$ corresponds to the elements $\{O_X(m) : m$ in $Z\}$ , we have that $\underline{\omega}_\Delta \cong O_X(m)$ for some $m$ , and we may apply Theorem (6.4).

<u>Proof of Theorem (6.4)</u>. The necessity of (1) is clear. Let

$$D_\Delta = \mathrm{Ext}_S^{n-d}(K[\Delta],S) ,$$

so that $\underline{\omega}_\Delta = \tilde{D}_\Delta$ . If $K[\Delta]$ is Gorenstein, then since $D_\Delta$ is a canonical module, $D_\Delta$ is a (graded) rank one projective over $K[\Delta]$ , and, hence, free, i.e., $D_\Delta \cong R(m)$ for some $m$ , where $R = K[\Delta]$ and $(m)$ indicates shift of grading:

$$[R(m)]_h = R_{m+h} .$$

But then $D_\Delta^{\tilde{r}} \cong R(m)^{\tilde{}} = O_\Delta(m)$ .

Now assume conditions (1) and (2). We must show that $K[\Delta]$ is Gorenstein. It suffices to show that the graded R-module $D_\Delta$ , which is a canonical module, is free (since $K[\Delta]$ is known already to be C-M), i.e., that $D_\Delta \cong R(m)$ for some positive integer $m$ (as graded R-modules).

Now, with $m$ as in condition (2), we know that $\tilde{D}_\Delta \cong O_\Delta(m) \cong R(m)^{\tilde{}}$ , and we know that $D_\Delta$ and $R(m)$ are both C-M modules of maximal dimension over $R$ . Hence, if $p = \oplus_{i\geq 1} R_i$ we have (since $\dim(\Delta) \geq 2$ )

that $\text{depth}_p(R(m)) = \text{depth}_p(D_\Delta) = \text{height}(p) = \dim(\Delta + 1) \geq 2 \Rightarrow H_p^0(E) =$ $H_p^1(E) = 0$ , where $E = D_\Delta$ or $R(m)$ . Hence, by the discussion on the bottom of p. 134 of [13], we have

$$D_\Delta \cong \bigoplus_{-\infty < t < \infty} \Gamma(X_\Delta, D_\Delta(t)\tilde{\ })$$

$$\cong \bigoplus_{-\infty < t < \infty} \Gamma(X_\Delta, R(m)(t)\tilde{\ })$$

$$(\text{for } \quad D_\Delta(t)\tilde{\ } \cong O_\Delta(m + t) \cong R(m)(t)\tilde{\ })$$

$$\cong R(m) \quad .$$

We next remark:

Proposition (6.6).    Propositions (2.3) and (2.4) hold with "Gorenstein" replacing "C-M" throughout.

The next few propositions give some explicit, useful, sufficient (and, sometimes, necessary) conditions for $K[\Delta]$ to be Gorenstein.

Theorem (6.7). Let $\Delta$ be a simplicial complex, $\dim(\Delta) \geq 1$ . The following conditions on $\Delta$ are equivalent:

(1)   $Z[\Delta]$ is Gorenstein.

(2)   $Z[\Delta]$ is C-M, and for each $(\dim(\Delta - 2))$-simplex $\sigma$ in $\Delta$ , $L_\Delta(\sigma)$ is either a circle, or a line with at most 3 vertices.

(3)   For every prime field $K$ of positive characteristic, $K[\Delta]$ is Gorenstein.

(4)  For every Gorenstein ring  K ,  K[Δ]  is Gorenstein.

Proof.  The equivalence of (1), (3) and (4) follows from (6.6).  (2)
is clearly necessary:  $Z[L_\Delta(\sigma)]$  will be Gorenstein, and  $L_\Delta(\sigma)$  is
1-dimensional.  Thus, the interesting fact is that (2) implies (1).
Thus, assume (2).  We must show that  Z[Δ]  is Gorenstein.  We use
induction on  dim(Δ) .  If  dim(Δ) = 1 ,  $\Delta = L_\Delta(\emptyset)$ .  ( ∅  is a
(-1)-simplex) is a circle, or line with at most 3 vertices, and we
are done.

    Now assume  dim(Δ) ≥ 2 .  The hypothesis of (2) passes to links
of vertices.  Hence, for each vertex  x ,  $K[L_\Delta(x)]$  is Gorenstein,
by the induction hypothesis.

    Thus, we may conclude from (6.5) that  Z[Δ]  is Gorenstein provided

$$H^1(\Delta;Z_2) = 0$$

(since  $U(Z) \cong Z_2$ ).  But  Z[Δ]  C-M implies  $Z_2[\Delta]$  C-M, which implies
$\tilde{H}^1(\Delta;Z_2) = H^1(\Delta;Z_2) = 0$  if  dim(Δ) ≥ 2 .

Corollary (6.8).  If  |Δ|  is a sphere,  K[Δ]  is Gorenstein.

    This answers a question of R. Stanley.  The special role of  $Z_2$
here is analogous to the special role  $Z_2$  plays in the discussion of
orientability of manifolds.

Proposition (6.9).  Let  K  be a field of  char(2) .  Suppose  dim(Δ) ≥ 1 .

Then  K[Δ]  is Gorenstein if and only if  K[Δ]  is C–M and for each

(dim(Δ − 2))–simplex  σ  in  Δ ,  $L_Δ(σ)$  is either a circle, or a line

with at most 3 vertices.

In this case  R[Δ]  will be Gorenstein for each Gorenstein  R  such

that the image of each odd integer in  R  is invertible.


Proof. For the first part we may assume  $K = Z_2$ .  The argument is then

the same as in (2) implies (1) in the proof of Theorem (6.7), except

that we conclude that

$$H^1(Δ;U(K)) = 0$$

here because  U(K)  is trivial!

Once we know that  $Z_2[Δ]$  is Gorenstein, we also know that  Q[Δ]

is Gorenstein, and in fact, that  L[Δ]  is Gorenstein for all fields of

characteristic 2 or 0 . The hypothesis on  R  implies that every field

R  maps into has characteristic 2 or 0 .


Proposition (6.10). Suppose  $\dim(Δ) \geq 1$ ,  Q[Δ]  is C–M, and for each

(dim(Δ − 2))–simplex  σ  in  Δ ,  $L_Δ(σ)$  is a circle or a line with at

most 3 vertices.

Suppose also that if  $Δ_0$  is either  Δ  or a link of a simplex in

Δ  and  $\dim(Δ_0) \geq 2$ , then  $H^1(Δ_0;Z_2) = 0$ .

Then for every Gorenstein ring  R  which contains  Q ,  R[Δ]  is

Gorenstein.

__Proof__.  R  maps only into fields of characteristic  0  and so it suffices

to show that  Q[Δ]  is Gorenstein.  It is enough to show that for some

prime  p ,  $Z_p$[Δ]  is Gorenstein.

Let  $Δ_0$  run through  Δ  and its links of dimension  $\geq 2$ .  There

are only finitely many primes  q  such that  $H^1(Δ_0;Z_2) \neq 0$  for some such

$Δ_0$ , since, by the C-M property for  Q[Δ] ,  $H^1(Δ_0;Q) = 0$  for any such

$Δ_0$ .  Let  $q_1,\ldots,q_t$  be all these primes.  By hypothesis, all the  $q_i$

are odd, and so there is a prime  p  in the arithmetic progression

$$\{(2h + 1)(q_1\ldots q_t) + 2 : h = 0,1,2,\ldots\} ,$$

in fact, infinitely many.  Hence we can choose  p  such that

$$q_i \nmid (p - 1) , \quad i = 1,\ldots,t .$$

We shall show that  $Z_p$[Δ]  is Gorenstein.  As before, using

induction on  dim(Δ) , the fact that the hypotheses pass to links, and

so forth, we come down, by  (6.5), to the problem of showing that

$$H^1(Δ_0;U(Z_p)) = 0 ,$$

for links  $Δ_0$  of dimension  $\geq 2$ , as above.  Since  $U(Z_p)$  has

order  p - 1 , it has a filtration by copies of  $Z_q$ , where  q  is not

in  $\{q_1,\ldots,q_t\}$  is prime, and the result follows.


Although it is possible to make a substantially finer analysis of

the Gorenstein property and related properties, we shall pursue the issue

no further here.

However, we do want to point out that there is a relationship between Serre-Grothendieck duality on $\mathrm{Proj}(K[\Delta])$ and Poincaré duality on $|\Delta|$ .

For the rest of this section, assume that $K$ is a field, and that $\Delta$ is a simplicial complex of pure dimension $d \geq 2$ . Assume also that each link of a vertex of $\Delta$ is C-M <u>but not that</u> $K[\Delta]$ <u>is</u> <u>C-M</u>.

<u>Theorem (6.11)</u>. Under the hypotheses above

(i)  $H^i(X_\Delta ; 0_\Delta) \cong H^i(\Delta ; K)$ , $1 \leq i < \dim(\Delta)$

(ii)  $H^i(X_\Delta ; 0_\Delta(m)) = 0$ , $m \neq 0$ , $1 \leq i < \dim(\Delta)$

(iii)  $H^d(X_\Delta ; 0_\Delta(m)) = 0$ , $m > 0$ , $d = \dim(\Delta)$ .

<u>Proof</u>. Let $R = K[\Delta]$ and $P = \oplus_{i \geq 1} R_i$ . One can reduce the characteristi 0 case to the characteristic $p$ case and in characteristic $p$ one may assume that $K$ is perfect or even $Z/pZ$ . We then deduce the result from Theorem 1.1 of [13], using the fact that

$$H^i_P(R) \cong [H^i(K^\cdot(x_0, \ldots, x_n ; R))]_0 ,$$

where $K^\cdot$ denotes the Koszul complex with the numbering reversed and the fact that

$$[H^i_P(R)]_m \cong H^{i-1}(X, 0_X(m)) , \quad i \geq 2 .$$

The vanishing results are immediate from Theorem (2.1). Now, for $1 \leq i < \dim(\Delta)$ we have

$$H^i(X_\Delta, O_\Delta) \cong [H^{i+1}(K^\bullet(x_0, \ldots, x_n; R))]_0$$
$$\cong H^{n-i}([K.(x_0, \ldots, x_n; R)]_0) \ .$$

The grading on $K^\bullet(\underline{x}; R)$ is such that a K-basis for $[K^{i+1}(\underline{x}; R)]_0$

consists of products of monomials of degree $i + 1$ and elements of the

free basis: hence, in $[K_{n-i}(x_0, \ldots, x_n; R)]_0$ we allow only monomials of

degree $i + 1$ . If we trace this restriction through our identification

of $\mathrm{Tor}^S_{n-i}(K, K[\Delta])$ ( $\cong H_{n-i}(K.(\underline{x}; R))$ ) with

$$\underset{\substack{-1 \leq q \leq i \\ T \subset V, \mathrm{card}(T+q)=i}}{\oplus} \tilde{H}^q(\Delta/T; K)$$

we see that the 0-th graded piece corresponds to the unique term for

which $T = \emptyset$ . (We are following the same argument as in [23].) Thus,

$$H^i(X_\Delta, O_\Delta) \cong \tilde{H}^i(\Delta : K) \cong \tilde{H}^i(\Delta; K) \ , \quad 1 \leq i < \dim(\Delta) \ ,$$

as required.


Discussion. We can now relate Serre-Grothendieck duality on

Proj(K[$\Delta$]) with Poincaré duality on $|\Delta|$ .

Suppose that $|\Delta|$ is a manifold, or even merely that $\Delta$ is of

pure dimension $d$ and that for each vertex $x$ , $K[L_\Delta(x)]$ is Gorenstein.

(In a manifold, all proper links are homology spheres and hence all

proper links are Gorenstein.) Suppose that $H^1(\Delta; U) = 0$ , where $U$ is

the units of the prime field of $K$ . (In fact, it is possible to allow

$K = Z$ , in which case $U$ is only $Z_2$ .)

The condition on the links of vertices implies that $\underline{\omega}_\Delta$ is locally free, while the condition on $H^1(\Delta;U)$ implies that $\underline{\omega}_\Delta \cong \mathcal{O}_\Delta(m)$ for some $m$.

Now if $F$ is any locally free sheaf on $X_\Delta$ we have

$$H^i(X,F)^* \cong H^{d-i}(X,\underline{\omega}_\Delta \otimes F^\vee) ,$$

(see (3.7)) where $^* = \operatorname{Hom}_K(\ ,K)$ and $^\vee = \operatorname{Hom}_{\mathcal{O}_\Delta}(\ ,\mathcal{O}_\Delta)$. In particular,

$$H^i(X,\mathcal{O}_\Delta)^* \cong H^{d-i}(X,\mathcal{O}_\Delta(m)) .$$

Thus, if $m \neq 0$, we obtain

$$H^i(X,\mathcal{O}_\Delta) = 0 , \qquad 1 \leq i < d$$

(and it is possible that $m \neq 0$).

If $m \neq 0$, we have

$$H^i(X,\mathcal{O}_\Delta)^* \cong H^{d-i}(X,\mathcal{O}_\Delta) ,$$

and, by (6.11), $H^i(X,\mathcal{O}_\Delta) \cong H^i(\Delta;K)$, $1 \leq i < d$. Thus, whether $m \neq 0$ or $m = 0$, we have

$$H_i(\Delta;K) \cong H^{d-i}(\Delta;K)^* , \qquad 1 \leq i \leq d - 1 ,$$

i.e., as claimed earlier, Serre-Grothendieck duality reduces to Poincaré duality.

§7. The Cohen-Macaulay Property for Ideals Generated by Monomials which

may Involve Higher Powers. In this section we sketch quickly a technique

for dealing with arbitrary ideals generated by monomials.

$$x_{i_1}^{h_1} \ldots x_{i_t}^{h_t}$$

in the indeterminates $x_0, \ldots, x_n$. For simplicity, we deal only with

fields.

Let K be a field, let $x_0, \ldots, x_n$ be indeterminates over K, and

let $S = K[x_0, \ldots, x_n]$. Let J be an ideal of S generated by monomials

in the x's. If f is another monomial in the x's, f is not in J,

we shall call $Rad(J:f)$ an **associated radical ideal** of J.

$Rad(J:1) = Rad(J)$ is always one of these: let $S/Rad(J) = K[\Delta] =$

$S/I_\Delta$, where $\Delta$ is the simplicial complex of supports of monomials

not in $Rad(J)$. $Rad(J)$ is always one of the associated radical ideals

of J, while the others correspond to certain subcomplexes of $\Delta$.

Hence, J has only finitely many associated radical ideals.

Theorem (7.1). With the above notation:

$$depth(S/J) = \min\{depth(S/I) : I \text{ is an}$$

$$\text{associated radical ideal of } J\}.$$

Hence, S/J is Cohen-Macaulay if and only if all the S/I are Cohen-

Macaulay of the same dimension.

Proof. One may reduce as in [13] to the case where K is a perfect field of characteristic $p > 0$. Let $^e(S/J)$ denote $S/J$ regarded as a module over itself via the e-th power of the Frobenius homomorphism. Then

$$\text{depth}_{S/J}(S/J) = \text{depth}_{S/J}{}^e(S/J) .$$

It suffices to show that for large $e$, $^e(S/J)$ is a direct sum of cyclic modules with radical annihilators such that the set of annihilators which occur is precisely the set of associated radical ideals of $J$. To see this, first note that, for any $e$, the monomials

$$g = x_0^{h_0} \cdots x_h^{h_n}$$

such that each $h_i < p^e$ and $g$ not in $J$ give rise to a direct sum decomposition of $^e(S/J)$ into cyclic modules. The annihilator of the cyclic module arising from a given $g$ is

$$\{r \text{ in } S/J : r^{p^e} g \text{ is in } J\} .$$

For fixed $g$ and large $e$ this is precisely

$$\text{Rad}(J:gR) .$$

Thus, every associated radical ideal occurs for $e$ sufficiently large.

Moreover, if $p^e$ exceeds all integers which occur as exponents in a set of generating monomials for $J$, then

$$r^{mp^e} g \text{ in } J , \quad m \geq 1 , \quad r \text{ a monomial} ,$$
$$\text{implies } r^{p^e} g \text{ is in } J ,$$

regardless of which monomial  g  is.  Thus, for large  e  all the

annihilators are (associated) radical and all the associated radical

ideals occur.

     Example.  Let  $I = (xu, xv, yv)$  in  $K[x, y, u, v]$ , which corresponds

to the simplicial complex

$$x \qquad y \qquad u \qquad v$$

    Thus,  $S/I$  is C—M.  However,  $S/I^2$  is not C—M, since

$$I^2 : xv = I + (yu) \quad \text{is radical}$$

$((xv)(yu) = (xy)(uv)$  in  $I^2$ ) , and  $I + (yu)$  corresponds to

$$x \qquad y \qquad u \qquad v$$

which is disconnected and, hence, not C—M.

Added in Proof.

a)  R. Stanley has written the author about the following new
results:  Let  $\Delta$  be a simplicial complex of pure dimension  d .  Let  K
be a field (all cohomology if taken with coefficients in  K ).

    i)  J. Munkres has shown that  $K[\Delta]$  is C-M if and only if for
all  x  in  $|\Delta|$ ,  $H^i(|\Delta|,|\Delta| - \{x\}) = H^i(|\Delta|) = 0$ ,  $i < d$ , and
has given a similar characterization of depth  $K[\Delta]$ .

    ii)  J. Munkres and R. Stanley have shown, independently, that
$K[\Delta_1]$  is Gorenstein if and only if  $\Delta_1$  is the join of a simplex
and a simplicial complex  $\Delta$  as above such that  $K[\Delta]$  is C-M and
for all  x  in  $|\Delta|$ ,  $H^d(|\Delta|,|\Delta| - \{x\}) \cong H^d(|\Delta|) \cong K$ .

b)  The present author has given, independently, more complicated
characterizations of the C-M property (and of depth) for  $K[\Delta]$  in terms
of purely topological properties of  $|\Delta|$ , along the lines indicated in
[8] (but with cohomology over  K  replacing homotopy).

c)  The author wishes to thank R. Stanley for correcting the
reference to a non-shellable sphere, and L. Billera for correcting the
inequalities in Cor. (5.3).

# REFERENCES

1.  Altman, A. and S. Kleiman.  Introduction to Grothendieck Duality
    Theory, Lecture Notes in Math. # 146, Springer, New York, 1970.

2.  Bass, H.  "On the ubiquity of Gorenstein rings," Math. Z. 82(1963),
    8-28.

3.  Brugesser, H. and P. Mani.  "Shellable decompositions of cells and
    spheres," Math. Scand. 29(1971), 197-205.

4.  Eagon, J. A. and M. Hochster.  "R-sequences and indeterminates,"
    Quart. J. Math. Oxford Ser. (2)25(1974), 61-71.

5.  Grothendieck, A. (notes by R. Hartshorne).  Local Cohomology,
    Lecture Notes in Math. # 41, Springer, Berlin, 1967.

6.  Hamann, E.  "On the R-invariance of  R[x] ," Thesis, Univ. of
    Minnesota, 1973, and J. of Alg. 35(1975), 1-16.

7.  Hochster, M.  "Cohen-Macaulay modules," Proc. Kansas Commutative
    Algebra Conference, Lecture Notes in Math. # 311, Springer, Berlin,
    1973, pp. 120-152.

8.  _____.  "Rings of invariants of tori, Cohen-Macaulay rings
    generated by monomials, and polytopes," Ann. of Math. 96(1972), 318-337.

9.  _____.  "Grassmannians and their Schubert subvarieties are
    arithmetically Cohen-Macaulay," J. of Alg. 25(1973), 40-57.

10. Hochster, M. and J. A. Eagon.  "Cohen-Macaulay rings, invariant
    theory, and the generic perfection of determinantal loci," Amer.
    J. Math. 93(1971), 1020-1058.

11. Hochster, M. and L. J. Ratliff, Jr. "Five theorems on Macaulay rings," Pacific J. Math. 44(1973), 147-172.

12. Hochster, M. and J. L. Roberts. "Rings of invariants of reductive groups acting on regular rings are Cohen-Macaulay," Adv. in Math. 13(1974), 115-175.

13. _____. "The purity of the Frobenius and local cohomology," Adv. in Math. 20(1976).

14. Hilton, P. J. and S. Wylie. Homology Theory, Cambridge University Press, London, 1960.

15. Kaplansky, I. "R-sequences and homological dimension," Nagoya Math. J. 20(1962), 195-199.

16. _____. Commutative Rings, Allyn and Bacon, Boston, 1970.

17. Klee, V. "The number of vertices of a convex polytope," Can. J. Math. 16(1964), 701-720.

18. Macaulay, F. S. "Some properties of enumeration in the theory of modular systems," Proc. London Math. Soc. 26(1927), 531-555.

19. Matijevic, J. R. and P. Roberts. "A conjecture of Nagata on graded Cohen-Macaulay rings," to appear.

20. McMullen, P. "The maximum numbers of faces of a convex polytope," Mathematika 17(1970), 179-184.

21. McMullen, P. and G. C. Shephard. Convex Polytopes and the Upper Bound Conjecture, London Math. Soc. Lecture Note Series 3, Cambridge Univ. Press, 1971.

22. Motzkin, T. S. "Comonotone curves and polyhedra, Abstract III," Bull. Amer. Math. Soc. 63(1957), 35.

23. Reisner, G. A. "Cohen-Macaulay quotients of polynomial rings," Thesis, University of Minnesota, 1974; _Adv. in Math._ 21(1976), 30-49.

24. Rudin, M. E. "An unshellable triangulation of a tetrahedron," _Bull. Amer. Math. Soc._ 64(1958), 90-91.

25. Siu, Y.-T. _Techniques of Extension of Analytic Objects_, Marcel Dekker, New York, 1974.

26. Spanier, E. H. _Algebraic Topology_, McGraw-Hill, New York, 1966.

27. Stanley, R. "The Upper Bound Conjecture and Cohen-Macaulay rings," _Studies in Applied Math._ 54(1975), 135-142.

28. Taylor, D. "Ideals generated by monomials in an R-sequence," Thesis, University of Chicago, 1966.

29. Traverso, C. "Seminormality and Picard group," _Ann. Scuola Norm. Sup. Pisa_ 24(3)(1970), 585-595.

30. Edwards, R. D. _Notices of the Amer. Math. Soc._ 22(1975), #A-334.

# EXAMPLES OF ARTINIAN, NON-NOETHERIAN RINGS

Lawrence S. Levy

University of Wisconsin

§1. **Introduction**. The purpose of this note is to convince the
reader that the left Artinian, left non-Noetherian rings (notation
**left** A **not** N **rings**) form a large class with some interesting
structure. For details we refer to [4].

Hopkin's well-known theorem states that a left A not N ring
cannot contain an identity element. (A more customary statement is
that, for rings with identity, left A implies left N .) In fact,
Hopkins proved that a left A not N ring cannot have either a left
or right identity ([2], 6.7). Trivial examples of left A not N
rings are given by either $Z(p^\infty)$ groups, multiplication being defined
to be zero; or direct sums of these with your favorite left A **and**
N ring with identity. To avoid this uninteresting situation, we will
focus on indecomposable left A and N rings, **indecomposable**
meaning that the ring is nonzero and is not the direct sum of two nonzero rings.

Let $D(R)$ be the largest divisible torsion subgroup of the
additive group $(R,+)$ of a left A ring R . Then

(1.1)                           $R \cdot D(R) = 0 = D(R) \cdot R$

(see [2], 72.3). This shows that every subgroup of $D(R)$ is a two-
sided ideal of $R$ . Since $R$ is left $A$ , this shows that $D(R)$ must
be _finitely decomposable_, that is, the direct sum of finitely many
indecomposable groups. The significance of $D(R)$ is given by the following theorem of L. Fuchs and T. Szele ([1], 73.3; or [3], 10.10).

**Theorem 1.2.** Let $R$ be a left **A** ring. Then $R$ is left $N$ if
and only if $D(R) = 0$ .

If we let $S(R) = R/D(R)$ , then Theorem 1.2 shows that $S(R)$ is left
$A$ _and_ $N$ . We will call $S(R)$ the _support ring_ of $R$ .

The significance of $S(R)$ in the structure of $R$ is given by
the result: _A given left_ A _and_ N _ring_ S _can support only finitely
many indecomposable, nonisomorphic left_ A _not_ N _rings_ R . (A
proof will be outlined in §2.) Furthermore, not every left $A$ and $N$
ring $S$ can support some indecomposable $R$ .

One restriction in the theorem of Szasz ([3], p.235, 10.17) that
every indecomposable left $A$ ring must have its underlying abelian
group primary or torsion-free. The torsion-free ones don't concern
us because, by Theorem 1.2, they are also left $N$ .

A second restriction on the structure of a support ring $S$ where
$0 \neq S = S(R)$ , with $R$ _indecomposable_ left $A$ not $N$ , is that $S$
cannot have a left or right identity element. (See §2.) This is a
bit stronger than Hopkins' theorem that $R$ cannot have a left or right

identity element.

   In §2 we sketch a structure theory of left  A  not  N  rings,
and then, in §3, we answer: What left  A  and  N  rings  S  can
support some indecomposable left  A  not  N  ring  R ?

§2.  **Reduction of**  A  **not**  N  **to**  A  **and**  N . Let  S  be a ring,  D
a divisible torsion group, and  $\delta : S \otimes_S S \to D$  any **additive** homomorphism.
We define the ring  $R = (S,D,\delta)$  by  $(R,+) = S \oplus D$  with multiplication given by

(2.1)    $(s_1 + d_1) \cdot (s_2 + d_2) = s_1 \cdot s_2 = (s_1 s_2) + \delta(s_1 \otimes s_2)$

where the dot indicates multiplication in  R , and multiplication in
S  is written  $s_1 s_2$ .

   As indicated by the notation,  D  will play the role of  D(R) ,
and  S  the role of  $S(R) \cong R/D(R)$ .  There is no choice in defining
(R,+)  to be  $S \oplus D$  because divisible groups are direct summands of
every abelian group containing them.  The absence of "cross products"
in (2.1) is due to the fact that  D(R)  is in the total annihilator of
R  (see (1.1)), and the  $\otimes_S$  merely expresses that multiplication is
associative.  After checking a few more details of this type, we
obtain the next result.

**Theorem 2.2.**  Every left  A  not  N  ring  R  equals some  $(S,D,\delta)$  with

$D = D(R)$   and

(1)  S  left  A  and  N , and  D  nonzero and finitely decomposable.

Conversely, if (1) holds, then  $R = (S,D,\delta)$  is left  A  not  N .

In order to use Theorem 2.2 effectively, we must know when rings
defined by it are isomorphic.

**Theorem 2.3.**  ("Uniqueness of  $\delta$ ").  The following assertions are
equivalent for a left  A  and  N  ring  S  and a divisible torsion group
D :

(1)  $(S,D,\delta_1) \cong (S,D,\delta_2)$ .

(2)  There is a group automorphism  $\theta$  of  D  **and** a ring automorphism  $\phi$  of  S  such that  $\delta_1 = \theta\delta_2(\phi \otimes \phi)$  on
$\ker(\mu_S : S \otimes_S S \to S^2)$ .

Before continuing with the theory, we pause to discuss the
**multiplication kernel**,  $\ker \mu_S$ , which occurs in (2) above and will be
a central consideration in virtually everything which follows.  First
of all, we have the following lemma.

**Lemma 2.4.**  If  S  has a left or right identity, then  $\ker \mu_S = 0$ .

For, suppose  e  is a left identity of  S .  Then every element of
$S \otimes_S S$  can be written in the form  $e \otimes s$ ; so  $0 = \mu_S(e \otimes s) = es = s$

implies  $e \otimes s = 0$ .

When  S  has no identity,  $\ker \mu_S$  can be nonzero.  However, one

of the most important technical results of the theory is the following

theorem.

**Theorem 2.5.**  The  $\ker \mu_S$  is a finite subgroup of  $S \otimes_S S$  when  S

is left  A  and  N .

To emphasize the role played by the multiplication kernel, we

state the following special case of Theorem 2.3.  (Take  $\theta$  and  $\phi$

equal to the identity on  D  and  S , respectively.)

(2.6)        $\delta_1 = \delta_2$  on  $\ker \mu_S$  implies  $(S,D,\delta_1) \cong (S,D,\delta_2)$ .

The next results give the first application of the theory.

**Theorem 2.7.**  Let  R  be an indecomposable left  A  not  N  ring with

$S(R) \neq 0$ .  Then  $S(R)$  cannot have a left or right identity element.

**Proof:**  By Theorem 2.2,  $R = (S,D,\delta)$  with  $D = D(R)$  and  $S \cong S(R)$ .

Suppose  S  has a left or right identity.  Then  $\ker \mu_S = 0$  by  Lemma

2.4 and therefore (2.6) shows

$$(S,D,\delta) \cong (S,D,0) = \underline{\text{ring}} \text{ direct sum } S \oplus D$$

contrary to indecomposability of  R .

Theorem 2.8. A given left A and N ring S can support only finitely many indecomposable, nonisomorphic left A not N rings R .

Proof (sketch): First we show that, given S and D , there are only finitely many nonisomorphic rings $(S,D,\delta)$ . By "Uniqueness of $\delta$ ," in the simplified form $(2.6)$, it suffices to show that there exist only finitely many additive maps: ker $\mu_S \to D$ ; and this follows from finiteness of ker $\mu_S$ (Theorem 2.5) and finite decomposability of D (Theorem 2.2).

The proof is then completed by showing that if D is too "big" (compared to ker $\mu_S$ ), then $(S,D,\delta)$ cannot be indecomposable.

§3. Examples. There is a "simplest possible" example of an indecomposable left A not N ring R with nontrivial multiplication. Let $D = Z(p^\infty)$ , and make the unique subgroup of order p into a ring $S^0$ by defining multiplication to be zero. We can define a second multiplication in this subgroup by identifying it with the integers modulo p . Denote this second product by $s_1 \otimes s_2$ .       Then $R = (S^0,D,inclusion$ has multiplication

$$(s_1 + d_1) \cdot (s_2 + d_2) = 0 + (s_1 \otimes s_2)$$

which amounts to saying: take the ordinary product $s_1 \otimes s_2$ , but put it into the second coordinate. ( R is indecomposable because its only minimal ideal is $R^2$ .)

This example is only barely nontrivial because $R^3 = 0$ . For a more satisfying family of examples we choose $S$ as close as possible to a ring with identity.

**Example-Type 1.** Let $S$ be any indecomposable, _idempotent_ ( $S = S^2$ ) left $A$ and $N$ ring for which $\ker \mu_S \neq 0$ . Then $S$ supports an indecomposable left $A$ not $N$ ring $R$ .

_Proof_: By Szasz's theorem, $(S,+)$ is p-primary for some $p$ . Let $D = Z(p^\infty)$ . Since $\ker \mu_S$ is a (finite) p-group, there is a nonzero additive homomorphism $\delta : \ker \mu_S \to D = Z(p^\infty)$ ; and since divisible groups are Z-injective, $\delta$ can be extended to an additive homomorphism: $S \otimes_S S \to D$ . Note that there will usually be many such extensions of $\delta$ ; but by (2.6), they all produce isomorphic rings $(S,D,\delta)$ .

We show that $R = (S,D,\delta)$ is indecomposable (and then we show how to obtain such an $S$ ).

Suppose $R = R_1 \oplus R_2$ with $R_i = (S_i, D_i, \delta_i) \neq 0$ . Then

$$R = (S_1 \oplus S_2, D_1 \oplus D_2, \delta_1 \oplus \delta_2) .$$

Since $D_1 \oplus D_2 = D(R) = Z(p^\infty)$ , an indecomposable group, we have either $D_1 = 0$ or $D_2 = 0$ . Say $D_2 = 0$ . Also $S_1 \oplus S_2 \cong S(R) \cong S$ is indecomposable so that $S_1$ or $S_2$ equals zero. If $S_2 = 0$ then $R_2 = 0$ , a contradiction. So $S_1 = 0$ . But then $\delta_1 \oplus \delta_2 = 0$ so "Uniqueness of $\delta$ " gives $0 = \theta\delta(\phi \otimes \phi)$ on $\ker \mu_S$ and hence $\delta = 0$ on ker $\mu_S$ , again a contradiction.  So  R  must be indecomposable as claimed.

But how does one obtain such an  S ?  Let  S  be **any** left  A  ring. Then  S/rad(S)  is a semisimple left  A  ring and hence has an identity element, which can be lifted to an idempotent  e  of  S  (called a **principal idempotent**).  We can use  e  to form the "Principal Pierce Decomposition" of  S :

(PPD)      $(S,+) = (1 - e)Se \oplus eSe \oplus eS(1 - e) \oplus (1 - e)S(1 - e)$    .

$$\underbrace{\phantom{(1-e)Se}}_{= A} \quad \underbrace{\phantom{eSe}}_{= U} \quad \underbrace{\phantom{eS(1-e)}}_{= B} \quad \underbrace{\phantom{(1-e)S(1-e)}}_{= N}$$

Here  1 - e  is used symbolically only, that is,  (1 - e)x  means x - ex .  Note that  U  is a ring with identity (the "unitary subring" of the PPD) and  A  and  B  are unitary U-modules.  Since  e  becomes the identity modulo  rad(S) ,  N  is contained in  rad(S) .  Hence  N  is a nilpotent ring.  Also,  A  and  B  are N-modules.  By squaring the PPD and using Nakayama's Lemma, we find that

(3.1)                          $S = S^2 \Leftrightarrow N = AB$ .

Moreover,

(3.2)                    $S = S^2 \Rightarrow \ker \mu_S = \ker(A \otimes_U B \to AB)$ .

Now we show that rings  S  with the desired properties abound ( $S = S^2$  is an indecomposable left  A  and  N  with  $\ker \mu_S \neq 0$ ).

Let  U  be any finite ring with identity which has a right ideal
A  which is <u>not projective</u>.  Then  $A_U$  is not flat, so there is a
left ideal  B  of  U  such that the map  $A \otimes_U B \to AB$  is not one-to-one.
Let  N = AB  (because of (3.1)), and make  $(S,+) = A \oplus U \oplus B \oplus N$
into a ring by declaring the right-hand side to be a PPD; that is,

$$S = \begin{bmatrix} U & B \\ A & N \end{bmatrix} \subseteq M_2(U) \; .$$

By (3.2),  $\ker \mu_S \neq 0$ ; and  S  can be shown to be indecomposable, so
we have our ring  S .  (Readers wanting a specific example can take  U
to be the integers  modulo  8  and  A = B = rad(U) .)

The rings  S  constructed above are all finite.  Infinite examples
can be obtained by a similar procedure, but more care must be exercised
to insure that  S  is left  A  (in which case, incidently,  S  cannot
be <u>right</u>  A ).


<u>Cross-Products Lemma</u>.  Let  $S = A \oplus B$  (ring direct sum) with  A  and
B  both left  A  and  N  and  (S,+)  primary.  Then

$$S \otimes_S S = (A \otimes_A A) \oplus (B \otimes_B B) \oplus \left[ \frac{A}{A^2} \otimes_Z \frac{B}{B^2} \right] \oplus \left[ \frac{B}{B^2} \otimes_Z \frac{A}{A^2} \right] \; .$$

Moreover, the last two terms on the right are nonzero if and only if
$A \neq A^2$  and  $B \neq B^2$ .

Example-Type 2.  If  $S = A \oplus B$  where  A  and  B  are any indecomposable
left  A  and  N  rings such that  $A \neq A^2$  and  $B \neq B^2$  (e.g.,  A  and
B  nilpotent) and  $(S,+)$  is p-primary, then  S  can support an indecomposable left  A  not  N  ring  R .

Sketch of Proof:  Note that the two right-most terms in the Cross-
Products Lemma belong to  $\ker \mu_S$ .  Hence it is easy to see that

$$(3.3) \qquad \ker \mu_S = \ker \mu_A \oplus \ker \mu_B \oplus \left( \frac{A}{A^2} \otimes_Z \frac{B}{B^2} \right) \oplus \left( \frac{B}{B^2} \otimes_Z \frac{A}{A^2} \right) .$$

Let  $\delta$  be the inclusion map of  $\ker \mu_S$  into its Z-injective hull  D ,
(since  $\ker \mu_S$  is finite,  D  is finitely decomposable) and extend  $\delta$
to a map:  $S \otimes_S S \to D$  using Z-injectivity of  D .

We claim  $R = (S,D,\delta)$  is indecomposable.  First note that  $\delta$  is
one-to-one on  $\ker \mu_S$ .  So, by "Uniqueness of  $\delta$ " (Theorem 2.3) this
will be true of any  $\delta'$  such that  $R \cong (S,D,\delta')$ .

Now suppose  $R = R_1 \oplus R_2$  with  $R_i = (S_i, D_i, \delta_i) \neq 0$ .  Then

$$(3.4) \qquad R = (S_1 \oplus S_2, D_1 \oplus D_2, \delta') \qquad (\text{for some } \delta') .$$

Note that  $S_1 \oplus S_2 \cong S(R) \cong S$ .  A Krull-Schmidt theorem can be proved
for Artinian rings.  (Remember:  our rings don't have identity elements.)
So we can suppose that  $S_1 \cong A$  and  $S_2 \cong B$ .  In particular, each
$S_i \neq S_i^2$ .

On the other hand, the way  $\delta'$  arises in (3.4) shows that it equals
$\delta_i$  on  $S_i \otimes S_i$  but equals zero on the "Cross-Product"

$(S_1/S_1^2) \otimes_Z (S_2/S_2^2)$ which, by the Cross-Products Lemma, is nonzero.
Therefore, (3.3) gives the contradiction that $\delta'$ is not one-to-one
on $\ker \mu_S$ .

**Example-Type 3.** If $S \neq S^2$ , $S$ is indecomposable left $A$ and $N$ ,
and $\ker \mu_S \neq 0$ , then $S$ supports some indecomposable left $A$ and $N$
ring $R$ .

The proof here is only a slight modification of the one above, so
will be omitted. The trouble here is that I don't know whether the
hypothesis $\ker \mu_S \neq 0$ follows from the other hypotheses.

The picture is completed by the surprising fact that any direct sum
of examples is again an example! More precisely:

**Example-Type 4.** If $S = \oplus_{i=1}^{n} S_i$ where $S$ is left $A$ and $N$ , $(S,+)$
is primary, and every $S_i$ supports an indecomposable left $A$ not $N$
ring $R$ , then $S$ supports a left $A$ not $N$ ring $R$ , too.

The proof combines the ideas of Types 1 and 2, but gets rather
technical, so we refer to ([4], 7.1) for the details.

Conversely, every $S$ which "works" must be of the types enumerated
above. So, except for uncertainty about which rings belong to Example-
Type 3, we have described all possible support rings $S$ of indecomposable
$R$ .

## REFERENCES

1.  Fuchs, L.  *Abelian Groups*, Pergamon Press, London, 1958.


2.  Hopkins, C.  "Rings with minimal condition for left ideals,"
    *Ann. of Math.* 40(1939), 712-730.


3.  Kertesz, A.  "Vorlesungen uber Artinische ringe," *Akademiai Kiadó*,
    Budapest, 1968.


4.  Levy, L. S.  "Artinian, non-Noetherian rings," to appear in *J. of*
    *Algebra*.

# THE CROSSED-PRODUCT PROBLEM

Murray Schacher

University of California, Los Angeles

§1. **Introduction.** Suppose $D$ is a division ring which is finite dimensional over its center $F$. It is well-known that $[D:F] = n^2$ for some integer $n$ called the **degree** of $D$. (For this and most other technical facts about finite dimensional division rings I will mention, a good reference is [5].) For any maximal subfield $L$ of $D$ we have $[L:F] = n = \sqrt{[D:F]}$, and standard arguments show that $D$ will have some maximal subfield which is separable over $F$ (even when char$(D) = p > 0$). The crossed-product conjecture asks, "Does $D$ have a maximal subfield which is a Galois extension of $F$?"

Perhaps we should pause to consider why one should want this conjecture to be true. Suppose then $D$ has a maximal subfield $L$ equipped with a group of automorphisms $G$ (necessarily of order $n$) so that $L^G = \{\alpha \in L \mid \sigma(\alpha) = \alpha, \text{ all } \sigma \text{ in } G\} = F$. In this case, we will say $D$ is a **crossed-product for** $G$. By the Skölem-Noether Theorem, for each $\sigma$ in $G$ there is a nonzero element $x_\sigma$ in $G$ (it is not unique!) so that the map $\sigma : L \to L$ agrees with inner automorphism by $x_\sigma$, i.e.,

$$x_\sigma \alpha x_\sigma^{-1} = \sigma(\alpha)$$

for all $\alpha$ in $L$. If $G = \{\sigma_1,\ldots,\sigma_n\}$, it is easily verified that the corresponding elements $\{x_{\sigma_1},\ldots,x_{\sigma_n}\}$ form a basis of $D$ as a left vector space over $L$, so

$$D \cong Lx_{\sigma_1} \oplus \cdots \oplus Lx_{\sigma_n}$$

and the additive structure of $D$ is determined. The multiplicative structure is obtained once we know how to evaluate the product $x_\sigma x_\tau$ for $\sigma$ and $\tau$ in $G$. One sees quickly that $x_\sigma x_\tau x_{\sigma\tau}^{-1}$ commutes with $L$, and so $f(\sigma,\tau) = x_\sigma x_\tau x_{\sigma\tau}^{-1}$ is in $L$ since $L$ is its own centralizer in $D$. The function $f : G \times G \to L^* = L - \{0\}$ is a 2-cocycle and a new choice of the $x_\sigma$'s will produce a cohomolgous cocycle. Thus, the structure of $D$ is determined completely by a Galois extension $L$, the Galois group $G$, and an element of the second cohomology group $H^2(G,L^*)$. This may not seem like much of a reduction to conventional ring-theorists, but cohomology groups like the one above have always been better understood than arbitrary division rings, and there exists elaborate machinery for dealing with them [see, for instance, [9]].

§2. The Amitsur Counter-example. The crossed-product conjecture is true when $F$ is an algebraic number field; in this case the group $G$ can always be taken to be cyclic. The proof requires a hefty chunk of class field theory (see [1]). The conjecture remains valid for general fields $F$ when $n$ is among the integers $2,3,4,6$ and $12$. Amitsur in [2] proved the conjecture is false in general. I would like to discuss his counter-examples and several questions suggested by them.

Let $K$ be a field and $n$ an integer. For $m \geq 2$, we construct $n$ generic $n \times n$ matrices $X_1, \ldots, X_n$; the entries of $X_i$ are $n^2$ distinct (but commuting) indeterminates; and, when $i \neq j$, we insist $X_i$ and $X_j$ involve distinct indeterminates. Let $R = K[X_1, \ldots, X_m]$ be the algebra generated by the $\{X_i\}$ over $K$. The basic properties of $R$ were first proved by Amitsur:

(1) $R$ is a noncommutative integral domain (with $m = 1$, $R$ would be commutative),

(2) $R$ satisfies all identities of $n \times n$ matrices over $K$,

(3) $R$ has a classical quotient ring $UD(K,n,m)$ which is a division ring of degree $n$ over its center.

I am using Jacobson's terminology; $UD$ stands for **universal division ring** (see [6]). A justification for this lofty title is provided by the following theorem.


**Theorem 1.** (Amitsur). (1) $UD(Q,n,m \geq 2)$ is a crossed-product for a group $G$ of order $n$ if and only if every division ring $D$ of degree $n$ and characteristic $0$ is a crossed-product for $G$ (where $Q$ denotes the rational numbers).

(2) If $Z_p(t)$ is the rational function field in one variable over the field of $p$ elements, ($p$ a prime), then $UD(Z_p(t),n,m \geq 2)$ is a crossed-product for a group $G$ of order $n$ if and only if every division ring $D$ of degree $n$ with $char(D) = p$ is a crossed-product for $G$.

§3.  <u>Noncrossed-Products</u>.  It is a short throw from Theorem 1 to the
construction of noncrossed-products; Amitsur did this by constructing
two "models:"

(A)  For any  $n \geq 2$ , a field  K  of iterated Laurent series with
char(K) = 0  and a central division ring  $D_1$  of degree  n  over  K  so
that all maximal subfields of  $D_1$  have an abelian Galois group with all
Sylow subgroups elementary abelian.

(B)  For any  $n \geq 2$ , a p-adic field  $Q_p$  and a central division
ring  $D_2$  of degree  n  over  $Q_p$  so that:  If  G  is an abelian group
which is the Galois group of a maximal subfield of  $D_2$ , then all Sylow
subgroups of  G  are cyclic or the direct product of  $Z_2$  with a cyclic
group.

The requirements in (A) and (B) are clearly incompatible if  $8|n$
or  $p^2|n$ ,  p  an odd prime.  Hence we have the next result.


<u>Theorem 2</u>.  (Amitsur).  UD(Q,n,m $\geq$ 2)  is not a crossed-product when
$8|n$  or  $p^2|n$ ,  p  an odd prime.


The first noncrossed-products in characteristic  p  were obtained,
using Amitsur's methods, by Schacher and Small.  They proved in [8] the
following theorem.


<u>Theorem 3</u>.  Suppose  p  is prime and  (n,p) = 1 .  Then  UD(Zp(t),n,m $\geq$ 2)
is not a crossed product when  $q^3|n$ ,  q  any prime.

The situation when  n  is not prime to  p  is more murky.  Some noncrossed-product results have been obtained in this case by Fein and Schacher in [4].  They proceed via the following definition.

Let  K  be a field and  G  a group of order  n .  We say  (K,G)  has Property A if every central division ring of degree  n  over  K  is a crossed-product for  G .

The groups  G  for which  (K,G)  has Property A are those which are elligible for making  $UD(K,n,m \geq 2)$  a crossed-product.  Most of our results concern the case where  K  is a global field of characteristic p , i.e., a finite extension of  $Z_p(t)$ .  In what follows we denote by  $|H|$  the order of a finite group  H .

Theorem 4.  (Fein-Schacher).  Suppose  K  is a global field with char(K) = p > 0  and  G  a group of order  n .  If  (K,G)  has Property A, then:

   (1)   The p-Sylow subgroup  P  of  G  is normal in  G , and, in
         fact,  G = PH ,  H  a subgroup of  G  and  $|H| = |G/P|$ .

   (2)   $H \cong Z_a \times Z_b$  where  $a|b$  and  K  contains a primitive a-th
         root of unity.

Jack Sonn has proved the converse of Theorem 4, so conditions (1) and (2) are necessary and sufficient for  (K,G)  to have Property A . (His work is, as yet, unpublished.)

Theorem 4 gives rise to a host of questions about noncrossed-products; some samples we state below.

Question 1. If $K$ is an infinite field with char$(K) = 3$ and $m \geq 2$, is UD$(K,G,m)$ a crossed-product for the symmetric group $S_3$ ?

Question 2. If $K$ is an infinite field with char$(K) = 2$ and $m \geq 2$, is UD$(K,8,m)$ a crossed-product for the dihedral group of order 8 ?

Question 3. If $K$ is a global field with char$(K) = p > 0$, and $m \geq 2$, is UD$(K,p^n,m)$ a crossed-product for every group of order $p^n$ ?

Question 3 is open even in the smallest case $n = 1$. In fact, not a single case of the crossed-product problem (in any characteristic) has been settled when $n$ is a prime $\geq 5$.

Theorem 4 is strong enough to produce some noncrossed-products in characteristic $p$ when $p \mid n$. To illustrate this, we have the following theorems.

Theorem 5. (Fein-Schacher). Let $K$ be an infinite field with char$(K) = p > 0$. Let $r$ be an integer which is prime to $p$ but divisible by the cube of a prime. Then, for any $a \geq 0$, $m \geq 2$, UD$(K,p^a r,m)$ is not a crossed-product.

Theorem 6. (Fein-Schacher). Let $K$ be a global field with $\text{char}(K) = p > 0$ and $r$ an integer satisfying:

  (a)  $p \nmid r$

  (b)  $q^2 | r$ ,  $q$  an odd prime

  (c)  $K$  does not contain a primitive q-th root of unity.

  Then, for any $a$ ,  $UD(K, p^a r, m \geq 2)$  is not a crossed-product.


  L. Risman claims a version of Theorem 6 with hypothesis (a) removed; I have not seen details of his proof.

  Another fascinating sequence of questions involve the center $Z$ of $UD(K,n,m \geq 2)$ . Procesi has proved in [7] that $Z$ is finitely generated over $K$ of transcendence degree $(m - 1)n^2 + 1$ . Further, he was able to prove $Z$ is purely transcendental over $K$ only in the case $n = 2$ . The case $n > 2$ is still open. A theorem of S. Bloch implies one of the following holds (perhaps both):

  (I)  The center of $UD(Q,n,m \geq 2)$ is not purely transcendental over $Q$
or

  (II) Every division ring of degree $n$ of characteristic 0 is

        similar (in the Brauer Group) to a tensor product of cyclic

        algebras, each of degree $n$ .

  We do not know which of (I) and (II) holds. (II) seems too fantastic to be true. A consequence of (II) would be that all central simple algebras in characteristic 0 have an abelian splitting field - this also is an open problem. Amitsur has proved (again in characteristic 0 ) that all division rings of degree $n$ have an abelian splitting field if and only if the universal one does.

Added in Proof:

A recent communication from David Saltman indicates that the results in
[10] show that Question 1 is true.

Concerning Question 3, Professor Amitsur also informs me that he and
Saltman have constructed algebras of degree $p^n$ ( $n \geq 2$ ) on
char = p which are crossed-products but not cyclic. Using these
methods Saltman has gone on to show the generic algebra of degree $p^n$
in char = p ( $n \geq 2$ ) is not a crossed-product. This would certainly
answer Questions 2 and 3 in the negative.

## REFERENCES

1. Albert, A. A.  Structure of Algebras, Amer. Math. Soc., New York, 1939.

2. Amitsur, S.  "On central division algebras," Israel J. Math. 12(1972), 408-420.

3. Bloch, S.  "Torsion algebraic cycles, $K_2$ , and Brauer groups of function fields," Bull. A.M.S. Vol. 80(1974), 941-945.

4. Fein, B. and M. Schacher.  "Galois groups and division algebras," J. Algebra, to appear.

5. Herstein, I. N.  Non-Commutative Rings,  Carus Monograph, 1968.

6. Jacobson, N.  PI-Algebras An Introduction, Springer-Verlag Lecture Notes #441, Berlin and New York, 1975.

7. Procesi, C.  Rings with Polynomial Identities, Marcel Dekker, New York, 1973.

8. Schacher, M. and L. Small.  "Noncrossed products in characteristic p ," J. Algebra 24(1973), 100-103.

9. Serre, J.-P.  Cohomolgie Galoisienne, Springer-Verlag, Berlin and New York, 1965.

10. Saltman, D.  "Splittings of cyclic p-algebras," Proc. A.M.S., to appear.

# GROWTH OF ALGEBRAS

Martha K. Smith

University of Texas

§<u>1</u>.  <u>Introduction</u>.  Let  A  be an algebra generated by the finite set

X  over the field  F .  Let  $X_n$  denote the subspace of  A  spanned by

all products of  n  or fewer elements of  X .  Define the <u>growth</u> <u>func-</u>

<u>tion</u>  $\gamma_X(n)$  <u>of</u>  A  <u>with</u> <u>respect</u> <u>to</u>  X  by

$$\gamma_X(n) = \dim_F X_n .$$

How does the behavior of  $\gamma_X(n)$  as  $n \to \infty$  affect the structure of

A  [8, p.375]?

   In order to talk about the growth of  A  independently of  X , it

will be convenient to make a few definitions.  All generating sets will

be assumed finite unless otherwise specified.  Say the nondecreasing

sequence  g(n)  <u>dominates</u> the nondecreasing sequence  f(n) , and write

f ⊂ g , if there exist positive numbers  a  and  b  such that  f(n) ≤

ag(bn) .  Note  ⊂  is clearly transitive.  Define  f ∿ g  if  f ⊂ g

and  g ⊂ f .  This is an equivalence relation compatible with  ⊂  and

(pointwise) multiplication of nondecreasing sequences.  Denote the

equivalence class of  f  by  [f] .  If  X  and  Y  each generate  A ,

then for suitable  c  and  c' ,  $X \subseteq Y_c$  and  $Y \subseteq X_{c'}$ , so

$\gamma_X(n) \leq \gamma_Y(cn)$ and $\gamma_Y(n) \leq \gamma_X(c'n)$ . Hence $[\gamma_X] = [\gamma_Y]$ . Thus define the **growth** $\gamma(A)$ to be the equivalence class $[\gamma_X]$ where X generates A . A has **exponential growth** if $\gamma(A) \supset [a^n]$ for some $a > 1$ ; otherwise A has **subexponential growth**. A has **polynomially bounded** (p.b. for short) growth if $\gamma(A) \subset [n^d]$ for some d and **polynomial growth** if $\gamma(A) = [n^d]$ for some integer d .

The number $\overline{\lim_{n \to \infty}} \dfrac{\log f(n)}{\log n}$ , if it exists, depends only on $[f]$ . Thus define the **Gelfand-Kirillov dimension** of A by

$$\text{GK dim A} = \overline{\lim_{n \to \infty}} \frac{\log \gamma_X(n)}{\log n} \qquad [7]$$

for any generating set S of A . GK dim A exists if and only if A has p.b. growth.

§**2.**  **Examples.**  The following examples will help give the flavor of the subject.

(a) If A is the free algebra with identity over F on a set S of cardinality d , then $\gamma_X(n) = \sum_{k=0}^{n} d^k$ , so A has exponential growth.

(b) If X is a set of d commuting indeterminates and A is the ring $F[X]$ of polynomials, then

$$\gamma_X(n) = \sum_{i=0}^{d} \binom{d}{i}\binom{n}{i} . \qquad [7],[32]$$

Hence  A  has polynomial growth and  GK dim A = d .

(c)  If  B  is a finitely generated subalgebra of  A , then
extending a set of generators of  B  to one of  A  shows that
$\gamma(B) \subset \gamma(A)$ .

(d)  If  A  and  B  are algebras with identity generated by  X
and  Y , respectively, then  $Z = X \otimes 1 \cup 1 \otimes Y$  generates  $A \otimes_F B$  and
$\gamma_Z(n) \subseteq \gamma_X(n)\gamma_Y(n)$ .

In particular, if  $A_n$  denotes the ring of  n × n  matrices over
A , then  $A_n = A \otimes F_n$ , so  $\gamma(A_n) \subset \gamma(A)\gamma(F_n) = \gamma(A)$  since  $[\gamma(F_n)] = [1]$ .
But by (c),  $\gamma(A) \subset \gamma(A_n)$ .  Hence  $\gamma(A) = \gamma(A_n)$ .

(e)  If  B  is a finitely generated subalgebra of  A  such that  A
is finitely generated as a B-module, then  A  may be embedded in  $B_n$  for
some  n .  It follows from (c) and (d) that  $\gamma(A) = \gamma(B)$ .

(f)  If  A  is commutative, then by [4], §3, there exists a set
$X = \{x_1, \ldots, x_n\}$  of algebraically independent elements such that  A  is
integral over  B = F[X] .  Since  A  is finitely generated as a module
over  B  [24, Theorem 17], it follows from (e) that  $\gamma(A) = \gamma(B)$ .  Hence
GK dim A = GK dim B = n  by (b).

Note that the Krull dimension  K dim A  of  A  equals  K dim B
[24, Theorem 48], which is also  n  [9, Theorem 9.2].  Thus  GK dim = K dim
for finitely generated commutative algebras.

(g)  The subspaces  $X_n$  form an ascending filtration of  A .  If
gr A  denotes the associated graded algebra, then  $\gamma(A) = \gamma(\text{gr } A)$ .  In
particular:

(i) Let $W_n$ denote the Weyl algebra of degree $n$ over $F$ ; that is, $W_n$ is generated over $F$ by $2n$ elements $p_1,\ldots,p_n$ ; $q_1,\ldots,q_n$ subject to the relations

$$p_i p_j - p_j p_i = q_i q_j - q_j q_i = 0$$
$$p_i q_j - q_j p_i = \delta_{ij} .$$

When $X = \{p_i\} \cup \{q_i\}$ , gr $W_n$ is the polynomial algebra in $2n$ indeterminates, so GK dim $W_n = 2n$ . (cf. [7], where the concept of GK dim was introduced precisely to distinguish between Weyl algebras of different degrees. See also [21], [22] and [28] for other work along these lines.)

(ii) If $L$ is a Lie algebra of dimension $n < \infty$ over $F$ , let $U(L)$ denote its universal enveloping algebra (cf. [17]). Choosing $X$ to be a basis for $L$ , gr $U(L)$ is the algebra of polynomials in $n$ indeterminates [17, p.166]. Hence GK dim $U(L) = n$ .

Here as in the commutative case there appears to be a connection with Krull dimension. Following [9], let $K$ dim $A$ denote the Krull dimension of $A$ in the sense of Gabriel-Rentschler and let $k$ dim $A$ denote the "little Krull dimension" (i.e., prime length). Then $k$ dim $A \subseteq K$ dim $A$ when the latter exists [9, Proposition 7.9]. Now $K$ dim $W_n = n$ [30, p.714] and $K$ dim $U(L) \leq n$ [5, Proposition 3.5.7]. Thus in each of these examples, both Krull dimensions are bounded by GK dim $A$ , but the case of $W_n$ shows that equality is not attained.

(h) Growth of groups may be considered in a manner analogous to that of algebras by defining $\gamma_X(n)$ to be the cardinality of $X_n$ where $X$ is a finite set of generators for the group $G$. $X$ also generates the group algebra $F[G]$ as an algebra, and $\gamma_X(n)$ in the group sense equals $\gamma_X(n)$ in the algebra sense. Growth of groups is of interest in studying curvature of Riemannian manifolds (cf. [26], [34]). It has been conjectured [2] that a group $G$ has subexponential growth if and only if $G$ is a finite extension of a nilpotent group. The latter condition is known to be sufficient, and necessary with the additional hypothesis that $G$ be solvable ([2], [11], [27], [34]). Bass' [2, Theorem 2] and Guivarc'h's [11, Theorem II.4] proofs of sufficiency are of particular interest since they show that a nilpotent-by-finite group in fact has polynomial growth. More specifically, if $G$ is a finite extension of the finitely generated nilpotent group $H$ with lower central series

$$H = H_1 \geq H_2 \geq \cdots \geq H_{p-1} = \{1\} ,$$

let $r_h$ denote the rank of $H_h/H_{h+1}$ and let

$$d = \sum_{h \geq 1} hr_h .$$

Then $\gamma(F[G]) = [n^d]$. Thus GK dim $F[G] = d$.

In line with the comments made in (f) and (g), it is worthwhile to note that by a result of P. F. Smith [33, Theorem 2.5],

K dim $F[G] = \sum_{h \geq 1} r_h \leq d$.

(j) Let $S$ denote the free algebra with identity on two generators

$\bar{x}$ and $\bar{y}$ . Let $A = S/S\bar{y}S\bar{y}S$ and let $x$ and $y$ denote the images

of $\bar{x}$ and $\bar{y}$ . Then with $X = \{x,y\}$ , $\gamma_X(n) = \dfrac{(n+1)(n+2)}{2}$ , so

$A$ has polynomial growth. However, the left ideals $Ayx^i$ form an

infinite direct sum, showing that $A$ does not have Krull dimension in

the sense of Gabriel-Rentschler [9, Proposition 1.4]. On the other

hand, it is easy to see that $k$ dim $A = 1$ .

(k) The Bass-Milnor-Wolf conjecture mentioned in (h) would imply

that the growth of a group is either exponential or polynomial. The

following class of examples shows that this is not the case for algebras.

Let $L$ be a finitely generated, infinite dimensional Lie algebra

with subexponential growth as a Lie algebra (e.g., the Lie algebra with

basis $x,y_1,y_2,\ldots$ such that $[x,y_i] = y_{i+1}$ , $[y_i,y_j] = 0$ ). Then

the growth of its universal enveloping algebra $U(L)$ is neither

exponential nor p.b. [32]. (The growth function of the specific example

mentioned, with $X = \{x,y\}$ , is $\gamma_X(n) = \sum\limits_{j=0}^{n} (n+1-j)P(n)$ , where

$P(n)$ is the number of partitions of $n$ . Note that $[P(n)] = [e^{\sqrt{n}}]$

[10, p.237]). These examples are domains [17, p.166], showing that

"intermediate" growth probably should not be considered an extremely

pathological condition.

§3. **A Necessary Condition.** So far, the only generally applicable

result giving a necessary condition for growth of a certain type is the

following result of G. Bergman, striking for its simplicity as well as

its content.

Proposition [3].  If  A  is an algebra with subexponential growth, then
any two elements of  A  have a common left (right) multiple.  In
particular, if  A  has no zero divisors, then  A  is a left and right
Ore ring.


Proof.  Let  X  generate  A ;  a,b  in  A .  Let  m  be such that  a,b  are
in  $X_m$ .  If  $Aa \cap Ab = \emptyset$ , then

$$\dim(X_i a + X_i b) = 2\gamma_X(i)$$

for every  i , so

$$\gamma_X(m + i) \geq 2\gamma_X(i) .$$

Consequently,

$$\gamma_X(km) \geq 2^{k-1}\gamma_X(m)$$

for every  k , and hence

$$[\gamma(A)] \supset [(2^{\frac{1}{m}})^n] .$$


§4.  Questions.  (1)  Example (b) shows that  GK dim  just measures
transcendence degree for commutative algebras.  The formula of example
(h) for  GK dim FG  may be regarded as a kind of weighted transcendence
degree.  One general question to be answered is if "weighted transcendence
degree" can be reasonably and precisely defined and linked with growth in

a more general setting.  One approach to transcendence is via the various
Krull dimensions.  Again, the Weyl algebras suggest a weighted transcendence degree:  GK dim $W_n$ = 2 K dim $W_n$ .  Specific questions for
starters along these lines:

(a)  Is  k dim A  always bounded by  GK dim A ?

(b)  Does a suitable well-behavedness condition (e.g.,
     Noetherian, existence of  K dim , finite uniform
     dimension) imply  K dim A $\leq$ GK dim A ?

(c)  Is there any exact formula for  K dim U(L)  in terms
     of the structure of  L ?

(2)  (a)  If  GK dim  exists, must it be an integer?

(b)  If  A  has p.b. growth, does it have polynomial growth?

(3)  For a given set  x  of generators, define  $\lambda_x(n) = \gamma_x(n) -$
     $\gamma_x(n - 1)$ .  $\lambda_x$  and  $\gamma_x$  have __generating functions__

$$F_x(t) = \sum_n \lambda_x(n)t^n$$

and

$$G_x(t) = \sum_n \gamma_x(n)t^n .$$

Note that  $G_x(t) = F_x(t) \sum_m t^m = \dfrac{F_x(t)}{1 - t}$ .  Generating functions proved
useful in studying growth of  U(L)  in [32].

(a)  What types of sequences occur as  $\lambda_x(n)$ 's?  Examples
     where  $\lambda_x(n)$  is not nondecreasing would be of interest.

(b)  (Bergman)  If  $G_x$  is rational for one set of generators of  A , is it for all?

(4)  (Bergman)  Let  A  be an algebra with subexponential growth
which is a domain.  Since  A  is an Ore domain by Bergman's result in
§3, it has a division ring of fractions [16, p.118].  Is the growth
of a finitely generated subalgebra of  D  influenced by that of  A ?
Some investigations along related lines have been done in [7], [21],
[22] and [28].

(5)  GK dim  distinguishes between various types of p.b. growth.
What types of subexponential but not p.b. growth can occur, and how can
they be characterized?


§5. Related Questions.  The study of growth in groups and its connection
with curvature of Riemannian manifolds was mentioned in §2(h).  The
growth of a group  G  also appears to be related to a certain property
of its  $L^1$  algebra.  The definition of growth function may be generalized
to locally compact, compactly generated groups by replacing cardinality
by Haar measure.  A Banach algebra with involution  x  is said to be
symmetric if  $1 + xx^*$  is invertible for every  x .  It has been conjectured [20] that for a locally compact group  G ,  $L^1(G)$  is symmetric
if and only if every compactly generated subgroup of  G  has subexponential
growth.  Sufficiency for a certain subalgebra was proved by J. Jenkins in
[20].  For other results related to this conjecture and similar analytic
questions connected with growth of groups, see [1], [6], [11]-[15], [18],
[19] and [31].

Added in Proof. Question 1 (a) has been answered affirmatively for Noetherian algebras and 2 has been answered negatively in W. Borho and H. Kraft, "Über die Gelfand-Kirillov-Dimension," Math. Ann. 220(1976), 1-24. Results relevant to questions 4 and 5 also appear in this paper. (Caution:  there appears to be an error in Satz 2.15.)  The bibliography contains additional references.

Another interesting property of the $L^1$ algebra of a group with p.b. growth can be found in Lemmas 6 and 7 of J. Dixmier, "Opérateurs de rang fini dans les représentations unitaires," Pub. Math. I.H.E.S. 6(1960), 305-317.

## REFERENCES

1.  Adel'son-Vel'skiĭ, G. M. and Yu. A. Šreĭder.  "The Banach mean on groups," Uspeki Mat. Nauk (N.S.) 12(1957), 131-136 (Russian).

2.  Bass, H. "The degree of polynomial growth of finitely generated nilpotent groups," Proc. Lond. Math. Soc. (3)25(1972), 603-614.

3.  Bergman, G.  Notes for Math 274, University of California, Berkeley, 1975.

4.  Bourbaki, N.  Elements de Mathématique, Algèbre commutative, Hermann, Paris, 1965.

5.  Dixmier, J.  Algèbres Enveloppantes, Gauthier-Villars, Paris, 1974.

6.  Emerson, W. R. and F. P. Greenleaf.  "Asymptotic behavior of products," Trans. Amer. Math. Soc. 145(1969), 171-204.

7.  Gelfand, I.M. and A. A. Kirillov.  "Sur les corp liés aux algèbres enveloppantes des algèbres de Lie," Pub. Math. I.H.E.S. 31(1966), 509-540.

8.  Gordon, R., Ed.  Ring Theory, Academic Press, New York, 1972.

9.  Gordon, R. and J. C. Robson.  Krull Dimension, Mem. Amer. Math. Soc. (133), 1973.

10.  Grosswald, E.  Topics from the Theory of Numbers, Macmillan, New York, 1966.

11.  Guivarc'h, Y.  "Croissance polynomiale et périodes des fonctions harmoniques," Bull. Soc. Math., France 101(1973), 333-379.

12.  Hulanicki, A.  "On the spectral radius of hermitian elements in
     group algebras," <u>Pac</u>. <u>J</u>. <u>Math</u>. 18(1966), 277-287.

13.  _____.  "On symmetry of group algebras of discrete nilpotent
     groups," <u>Studia</u> <u>Math</u>. 35(1970), 207-219.

14.  _____.  "On positive functionals on a group algebra multipli-
     cative on a subalgebra," <u>Studia</u> <u>Math</u>. 37(1971), 163-171.

15.  _____.  "On the spectrum of convolution operators on groups
     with polynomial growth," <u>Invent</u>. <u>Math</u>. 17(1972), 135-142.

16.  Jacobson, N.  <u>Theory of Rings</u>, American Mathematical Society,
     Providence, 1943.

17.  _____.  <u>Lie Algebras</u>, Interscience, New York, 1962.

18.  Jenkins, J. W.  "Growth of connected locally compact groups," <u>J</u>.
     <u>Funct</u>. <u>Anal</u>. 12(1973), 113-127.

19.  _____.  "Folner's condition for exponentially bounded groups,'
     <u>Math</u>. <u>Scand</u>. 35(1974), 165-174.

20.  _____.  "Representations of exponentially bounded groups,"
     <u>Amer</u>. <u>J</u>. <u>Math</u>. 98(1976), 29-38.

21.  Joseph, A.  "Gel'fand-Kirillov dimension for algebras associated with
     the Weyl algebra," Ann. Inst. H. Poincaré, Série A 17(1973), 325-336.

22.  _____.  "Symplectic structure in the enveloping algebra of a
     Lie algebra," <u>Bull</u>. <u>Soc</u>. <u>Math</u>., <u>France</u> 102(1974), 75-83.

23.  Justin, J.  "Groupes et semi-groupes à croissance linéaire," <u>C</u>.<u>R</u>.
     <u>Acad</u>. <u>Sci</u>., <u>Paris</u>, Série A 273(1971), 212-214.

24.  Kaplansky, I.  <u>Commutative Rings</u>, Allyn and Bacon, Boston, 1970.

25.  Milnor, J.  "Problem 5603," _Amer_. _Math_. _Monthly_ 75(1968), 685-686.

26.  _____.  "A note on curvature and the fundamental group," _J_.
     _Diff_. _Geom_. 2(1968), 1-7.

27.  _____.  "Growth of finitely generated solvable groups," _J_. _Diff_.
     _Geom_. 2(1968), 447-449.

28.  Nghiêm, X. H.  "Sur certains sous-corps commutatifs du corps
     enveloppant d'une algèbre de Lie résoluble," _Bull_. _Sc_. _Math_., Série
     2,96(1972), 111-128.

29.  Nouazé, Y. and P. Gabriel.  "Ideaux premiers de l'algèbre enveloppante
     d'une algèbre de Lie nilpotente," _J_. _Alg_. 6(1967), 77-99.

30.  Rentschler, R. and P. Gabriel.  "Sur la dimension des anneaux et
     ensembles ordonnés," _C_._R_. _Acad_. _Sci_., _Paris_, Série A 265(1967),
     712-715.

31.  Rosenblatt, J. M.  "Invariant measures and growth conditions," _Trans_.
     _Amer_. _Math_. _Soc_. 193(1974), 33-53.

32.  Smith, M. K.  "Universal enveloping algebras with subexponential
     but not polynomially bounded growth," _Proc_. _Amer_. _Math_. _Soc_., to appear.

33.  Smith, P. F.  "On the dimension of group rings," _Proc_. _Lond_. _Math_.
     _Soc_. 25(1972), 288-302.

34.  Wolf, J.  "Growth of finitely generated solvable groups and curvature
     of Riemannian manifolds," _J_. _Diff_. _Geom_. 2(1968), 421-446.

# THE ZERO DIVISOR CONJECTURE

Robert L. Snider

Virginia Polytechnic Institute and State University

§**1.** **Introduction.** Let $G$ be a group and $1 \neq g$ of finite order in $G$ . If the order of $g$ is $n$ , then in the group ring $F[G]$ , we have $(1 - g)(1 + g + \cdots + g^{n-1}) = 0$ . In other words, if $G$ has nontrivial elements of finite order, the group ring $F[G]$ has nontrivial zero divisors. On the other hand, if $G$ is torsion-free, then the conjecture is that the group ring $F[G]$ has no zero divisors.

The first thing one notices is that if $G$ is (fully) ordered, there are no zero divisors. This observation immediately takes care of abelian groups, nilpotent groups, and free groups.

In 1940, G. Higman invented a class of groups for which the ordered proof works [6]. This is the class of locally indicable groups. A group is **locally indicable** if each finitely generated subgroup has an infinite cyclic homomorphic image.

This is essentially all that was known until 1974, when Formanek proved the conjecture for supersolvable groups [4]. His proof uses essentially the fact that each supersolvable group has either $Z$ or the free product $Z_2 * Z_2$ as a homomorphic image.

Finally this past year, there has been a burst of activity. First of all, Kenneth Brown proved the conjecture when  G  is abelian-by-finite ( G  has an abelian subgroup of finite index) and the characteristic of the field is zero [1].  This was immediately followed by a proof in the case  G  is polycyclic-by-finite by Farkas and Snider in characteristic zero [2].  About the same time Jacques and Tekla Lewin proved the conjecture for one-relator groups [7].

§2.  **Polycyclic-by-Finite**.  I will give a sketch of the proof when  G  is polycyclic-by-finite.

**Theorem** [2].  If  G  is polycyclic-by-finite and the characteristic of F  if zero, then  F[G]  has no zero divisors.

We first recall that a group is **polycyclic** if it has a (finite) normal series with cyclic factors.  A group is **polycyclic-by-finite** if it has a polycyclic subgroup of finite index.

Let  G  be a torsion-free polycyclic-by-finite group.  First F[G]  is at least a prime ring by a theorem of Connell.  Then  F[G]  is Noetherian by repeated application of a noncommutative version of the Hilbert basis theorem.  Therefore, by the Goldie theorems,  F[G]  has a classical quotient ring which is simple Artin and hence isomorphic to $M_n(D)$ , a matrix ring over a division ring.

Now  G  has finite cohomological dimension  [5] and hence  F[G] has finite global dimension.  Let  I  be a right ideal of  F[G] .  Pick a finitely generated projective resolution for $I$ , say

$$0 \to P_n \to \cdots \to P_1 \to P_0 \to I \to 0 .$$

We then have

$$0 \to P_n \otimes_{F[G]} M_n(D) \to \cdots \to P_0 \otimes_{F[G]} M_n(D) \to I \otimes_{F[G]} M_n(D) \to 0$$

is exact. If one could show each $P_i \otimes M_n(D)$ was a free $M_n(D)$-module, then using additivity of dimensions over $D$ , one has $I \otimes M_n(D)$ is free. Now $I \otimes M_n(D) \cong IM_n(D)$ and hence $IM_n(D) = M_n(D)$ or $0$ . Therefore $n = 1$ and $F[G]$ is a domain.

We are reduced then to measuring the size of a projective module over $F[G]$ . If $P$ is finitely generated projective, then $P = E(F[G]^m)$ where $E$ is an idempotent in $M_m F[G]$ . To measure the size of a projective, one is immediately led to consider traces. In the group ring, one has a trace defined by $tr(\Sigma r_g g) = r_1$ . This can be extended to $M_m(F[G])$ by $tr(A) = \sum_{i=1}^{m} tr(a_{ii})$ . An argument of Formanek tells us that $tr(E)$ is an integer [3]. Therefore $tr(E)$ is the same as the trace of a free module. This is not enough as the trace function may not be extendable to $M_n(D)$ .

$G$ has a poly-infinite-cyclic subgroup $H$ of finite index. Projective modules over $F[H]$ are stably free. This follows from a noncommutative version of the theorem that $K_0(R) \cong K_0(R[x])$ . Also $F[H]$ has no zero divisors as $F[H]$ is constructed by a series of noncommutative polynomial extensions. (Also $H$ is locally indicable.)

Let $t = [G:H]$ . If $A$ is in $M_m(F[G])$ , then we can regard $A$

as in $M_{mt}(F[G])$ and we see that $tr_H(A) = t \ tr_G(A)$ .

Now $P$ is also projective as an $F[H]$-module since $F[G]$ is a free $F[H]$-module. Since $P$ is stably free as an $F[H]$-module, we have $P \oplus F[G]^r \cong F[H]^s$ . Taking traces of both sides we see that $t \ tr_G(P) + rt = s$ . If we tensor both sides with $Q$ , the quotient division ring of $F[H]$ , we obtain $(P \otimes_{F[H]} Q) \oplus (F[G]^r \otimes_{F[H]} Q) \cong Q^s$ . Now $F[G] \otimes_{F[H]} Q \cong M_n(D)$ and $P \otimes_{F[H]} Q \cong P \otimes_{F[G]} M_n(D)$ . $(P \otimes Q) \oplus M_n(D)^r \cong Q^s$ . Also $\dim_Q M_n(D) = t$ . Comparing dimensions, we see that $P \otimes Q$ has dimension $s-tr$ . But $t \ tr(P) + rt = s$ and hence $t|s$ . Therefore $P \otimes_{F[H]} Q \cong P \otimes_{F[G]} M_n(D)$ is a free $M_n(D)$-module.


§3.  **Concluding Remarks**.  Where does one go from here?  An obvious case to try is polycyclic-by-finite or even abelian-by-finite in characteristic $p > 0$ .  Another is the solvable case in characteristic 0  or perhaps solvable groups of finite cohomological dimension.

One-relator groups have cohomological dimension 2 .  It might be worth trying to do something with groups of finite cohomological dimension.


Editors Note:  A student of J.E. Roseblade has recently shown that if  G is Abelian-by-finite and the characteristic of  F  is  $p > 0$ , then  F[G] has no zero divisors.

REFERENCES

1.  Brown, K. A.  "Zero-divisors in group rings,"  to appear.

2.  Farkas, D. R. and R. L. Snider,  " $K_0$  and Noetherian group rings,"
    J. Algebra,  to appear.

3.  Formanek, E.  "Idempotents in Noetherian group rings," Canad. J.
    Math. 15(1973), 366-369.

4.  Formanek, E.  "The zero divisor question for supersolvable groups,"
    Bull. Austral. Math. Soc. 9(1973), 67-71.

5.  Gruenberg, K.  Cohomological Topics in Group Theory, Springer
    Lecture Notes No. 143.

6.  Higman, G.  "The units of group rings," Proc. London Math. Soc. 46(2)
    (1940), 231-248.

7.  Lewin, J. and T. Lewin.  "The group algebra of a torsion-free one-
    relator group can be embedded in a field," Bull. A.M.S. 81(1975),
    947-949.

PRIME IDEAL STRUCTURE IN NOETHERIAN RINGS

Roger Wiegand

University of Nebraska

§1. **Introduction.** I want to discuss some problems centered around the following question: Which partially ordered sets (posets) are order-isomorphic to spec(R) , the set of prime ideals of a suitable Noetherian (commutative) ring R ? This problem has its roots in algebraic geometry, since the prime ideals in $k[X_1,\ldots,X_n]$ are in one-one order-reversing correspondence with algebraic varieties in affine n-space $k^n$ . (Here k is an algebraically closed field, and the correspondence is given by $p \to \{x \in k^n \mid f(x) = 0$ for every $f \in p\}$ . The fact that this is a one-one correspondence is Hilbert's Nullstellensatz.) Many geometric results, particularly those of a dimension-theoretic nature, therefore amount to statements about prime ideal structure. (A good example is the theorem on the dimension of the intersection of two algebraic varieties, [13, Ch. VII, Theorem 27'].)

The prime ideal structure of the rings one studies in algebraic geometry is much tidier than that of more general Noetherian rings. Still, even if one were interested only in polynomials over a field, it would be important to understand the general situation. For one

thing, in proving theorems about "nice" rings, one can rarely work
entirely within the context of nice rings.  Inductive arguments
often involve passing to related rings that are no longer nice.  An
excellent example of this phenomenon is the recent theorem of Eisenbud
and Evans [1], that every algebraic set in n-space can be defined by
n  equations.  This is, essentially, a statement about prime ideals in
$k[X_1,\ldots,X_n]$ .  The key to the proof, however, was to prove the analogous statement in any Noetherian ring of the form  $R[X]$ .

§2.  **The Spectrum of an Arbitrary Commutative Ring**.  For non-Noetherian
rings the Zariski topology on  spec(R)  carries more information than the
partial ordering.  We recall the definitions:  If  I  is an ideal of  R ,
let  $V(I) = \{p \mid p \supseteq I\}$  and  $D(I) = spec(R) - V(I)$ .  The sets  $D(I)$
are the open sets in the Zariski topology.  This topology is sensitive
to the partial ordering, since  $p \subseteq q$  if and only if  $q \in \{p\}^-$ .

In particular,  spec(R)  satisfies the  $T_0$  separation axiom.  Now
it is easy to see that an open set  U  is quasi-compact if and only if
$U = D(I)$  for some finitely generated ideal  I .  Therefore  spec(R)
is quasi-compact and has an open basis consisting of quasi-compact open
sets; and the intersection of two quasi-compact open sets is again quasi-
compact.  There is one more important property of  spec(R) , again
easily verified:  Every irreducible closed set is the closure of a
singleton.  (A closed set is  **irreducible**  if it is  nonempty  and is not
the union of two closed proper subsets.)  In [3], Hochster defined a

topological space with these five properties ( $T_0$ , quasi-compact, etc.)
to be **spectral**, and he proved the following remarkable theorem.

**Theorem 1.**  Every spectral space is homeomorphic to  spec(R)  for
some commutative ring  R .

    This theorem, although extremely useful, doesn't seem to provide
a workable criterion for a given poset to be order-isomorphic to some
spec(R) .  The problem is that there may be lots of spectral topologies
that induce the same partial ordering.  (The partial ordering induced
by a  $T_0$  topology is:  $x \leq y$  if and only if  $y \in \{x\}^- .$ )  Take, for
example, the poset  $X = \{x, y_1, y_2, \ldots\}$  with order relations  $x < y_n$
for every  n .  One spectral topology on  X  is obtained by taking the
proper closed sets to be the finite subsets of  $Y = \{y_1, y_2, \ldots\}$ .
(This, of course, is the topology on  spec(Z)  of the integers  Z .)  But
one can enrich the topology without changing the ordering, by declaring any
subset of  Y  containing  $y_1$  to be closed, for example.  A complete catalog
of the spectral topologies on this poset may be found in [7].

    For Noetherian rings the distinction between topology and order
evaporates.  To see this, we note that if  R  is a Noetherian ring
then  spec(R)  is a Noetherian space, that is, it has the descending
chain condition on closed sets.  It follows that every closed set is
a finite union of irreducible closed sets; that is, the closed sets are
the finite unions of sets of the form  $\{x\}^- = \{y \mid y \geq x\}$ .  Thus the
partial ordering on the spectrum determines the topology.

These remarks give some specific information about the ordering of the primes in a Noetherian ring. More generally, suppose spec(R) is a Noetherian topological space and let U be order-isomorphic to spec(R) . Then:

(1)  U satisfies the ascending chain condition.

(2)  U has only finitely many minimal elements.

(3)  For each x,y ∈ U there are only finitely many elements
     z minimal with respect to x ≤ z ≥ y .

Also, by Zorn's lemma:

(4)  Every non-empty chain in U has a greatest lower bound.

Conversely, if U is any poset satisfying (1)-(4), we can topologize U by taking, as a closed subbase, the sets $\{x\}^- = \{y \mid y \geq x\}$ This turns out to be a Noetherian spectral topology, compatible with the given partial ordering. (See [12] for details.) By Hochster's theorem, there is a ring whose spectrum is order-isomorphic to U .

§3. **The Principal Ideal Theorem.** I doubt that there are any order-theoretic properties of the prime spectrum of a Noetherian ring, other than those deducible from (1)-(4) above, that one can prove without invoking Krull's principal ideal theorem. This theorem, published in 1926, [6], is the cornerstone of dimension theory in Noetherian rings. In order to state

and apply it we will need some terminology about posets; to avoid inter-
ruptions later, we will get it over with here.

Let $x < y$ in a poset. A **chain from** $x$ **to** $y$ is a sequence
$x = x_0 < x_1 < \cdots < x_n = y$ . The integer $n$ is the **length** of the chain.
The **height of** $y$ **over** $x$ , $ht(y/x)$ , is the supremum of lengths of
chains from $x$ to $y$ . The height and coheight of an element $z$ are
given by $ht(z) = \sup\{ht(z/x) \mid x < z\}$ and $coht(z) = \sup\{ht(y/z) \mid y > z\}$ .
The **dimension** $\dim X$ of a poset $X$ is $\sup\{ht(z) \mid z \in X\} =$
$\sup\{coht(z) \mid z \in X\}$ .

**Theorem 2.** (Principal Ideal Theorem). In a Noetherian ring, every
minimal prime of a principal ideal has height $\leq 1$ . (A minimal prime
of an ideal $I$ is just a minimal element of $V(I)$ .)

**Corollary**. In a Noetherian ring, every prime ideal is the union of
prime ideals of height $\leq 1$ .

We will see that this corollary cannot be formulated purely in terms
of the partial ordering on $spec(R)$ . Still, it allows us to enlarge
our list of axioms for the spectrum of a Noetherian ring. From now on,
U will denote a partially ordered set isomorphic to $spec(R)$ for some
Noetherian ring $R$ ; and any topological remarks will refer to the
topology induced by the partial ordering. Since a prime ideal cannot
be a finite union of smaller primes [5, Theorem 81], we deduce

(5)  If  dim U ≥ 2  then  U  has infinitely many elements of

     height  1 .

Thus, a 3-element totally ordered set cannot be the spectrum of

a Noetherian ring.  Similarly,

(6)  Let  x,y ∈ U  with  ht(x) ≥ 1 .  If  z ≤ y  whenever  z ≤ x

     and  ht(z) ≤ 1 , then  x ≤ y .

This rules out the following poset:

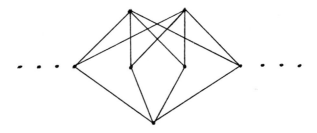

Of course, one can apply the principal ideal theorem to homomorphic

images of  R  to prove the relative versions of (5) and (6) in every

closed subset of  U .

Here is an example to show that neither the principal ideal theorem

nor its corollary can be stated in terms of the partial ordering (or

topology) on  spec(R) .


Example.  A Noetherian local domain  R  and a (non-Noetherian) domain

S  such that  spec(R)  and  spec(S)  are homeomorphic, but the maximal
ideal of  S  is not a union of height  1  primes.  Let  R  be any 2-
dimensional Noetherian local domain, say,  k[[x,y]] .  Apply Hochster's
construction [3] to build a ring  S  such that  spec(S) $\cong$ spec(R) ,
and let  M  be the maximal ideal of  S .  Since  spec(S)  is Noetherian
there are elements  $f_1,\dots,f_n \in S$  such that  $\{M\} = V(f_1,\dots,f_n)$ .  But
all the rings of [3] have the property:  $V(f_1,\dots,f_n) = V(f)$  for suitable
f .  (See the proof of [3, Theorem 4].)  Thus  $\{M\} = V(f)$ , and  f  is
not in any height  1  prime.

The next theorem, also published in Krull's 1926 paper, [6], is a
fairly easy consequence of the principal ideal theorem.


Theorem 3.  (Krull's Height Theorem).  In a Noetherian ring, a prime  p
has height  $\leq n$  if and only if  p  is minimal over some n-generator
ideal.


The subtler consequences of Theorem 3 will be deferred to the next
section; here we record only the obvious:

(7)   Each element of  U  has finite height.  In particular,  U
has the descending chain condition.

I should remark that Hochster [4] has formulated an order-theoretic
consequence of the height theorem, from which (5), (6), (7) and hence
(4), may be deduced easily:

(8)  U  has a closed base  $V$  with the property:  For every closed

subset  W  and every  $x \in W$,  $ht_W(x) \leq n$  if and only if there

exist  $V_1, \ldots, V_n \in V$  such that  $x$  is a minimal element of

$W \cap V_1 \cap \cdots \cap V_n$.  ( $ht_W(x)$  is the height of  $x$  in the

poset  W .)

It is conceivable that (8) is the best possible theorem about the
spectrum that is deducible from the height theorem.  To be precise:  Let
X  be any poset satisfying (1)-(3) and (8).  Is there a ring  S  such
that  spec(S)  is homeomorphic to  X , and  S  and all its homomorphic
images satisfy the conclusion of the Krull height theorem?

§4.  **Saturated Chains**.  We say  y  is a **cover** of  x  (in a poset) provided  $ht(y/x) = 1$.  A chain  $x_0 < \cdots < x_n$  is **saturated** provided
each  $x_{i+1}$  covers  $x_i$.  The **saturated chain condition** (on a
Noetherian ring  R ) is this:  For each pair of primes  $p \subset q$, all
saturated chains from  p  to  q  have the same length, namely,  $ht(q/p)$.
Rings with this property are sometimes called **catenary** rings.  Every
ring  $k[X_1, \ldots, X_n]$,  k  a field, is catenary, [13, Chapter VIII,
Theorem 20], as is every complete local domain [10, 34.4].  For a long
time it was unknown whether all Noetherian rings were catenary.  In
1956, Nagata [10] settled the question by producing a Noetherian catenary
ring  R  such that  R[X]  is **not** catenary.  Still, it is important to
know just how flagrantly the saturated chain condition can be violated.
The recent literature contains a few results that prevent "arbitrarily

bad" behavior:

(9)   Let  x ∈ U  with  ht(x) = n .  Then all but finitely many

covers of  x  have height  n + 1 .  (McAdam [8].)

(10)  Assume  U  has a unique maximal element and a unique minimal

element.  (That is,  U  is the spectrum of a Noetherian local

domain.)  Let  x ∈ U  with  ht(x) > 0  and  coht(x) = c > 0 .

Then, for each  i = 0,...,c - 1 , there are infinitely many

y ∈ U  such that  ht(y) = h + i  and  coht(y) = c - i .

(See [8].)

McAdam [8] has extended (10) to include semi-local domains.  Notice
that (10) is a considerable sharpening of (5).

The subtleties associated with failure of the saturated chain
condition seem to be the biggest obstacle to characterizing the spectra
of Noetherian rings.  Even formulating a reasonable conjecture appears
hopeless.  So, carefully avoiding these subtleties, we ask:

Questions.  (a)  Suppose  X  is a poset having the saturated chain condition and satisfying (1), (2), (3) and (8).  Is  X  order-isomorphic
to the prime spectrum of a Noetherian ring?

   (b)  Suppose  dim X ≤ 2  and  X  satisfies (1), (2), (3) and (8).
Is  X  order-isomorphic to the spectrum of a Noetherian ring?

A third question deals with the feeling that all theorems about the spectrum really use only the principal ideal theorem.


(c)  Let  S  be a commutative ring with Noetherian spectrum, and
assume that  S/p  satisfies the conclusion of the principal ideal
theorem for every prime  p .  Is there a Noetherian ring  R  such
that  $spec(R) \cong spec(S)$ ?


Most likely, the answer is "no"; however, a counter-example might
well lead to a new and useful theorem about Noetherian rings.  Question (c) is probably at odds with the question raised at the end of
§3, in view of an unpublished example of Ray Heitmann's, which appears to show that conditions (1), (2), (3) and (8) on a poset do not
imply (10).  (He has also shown, incidentally, that they do imply (9).)
Finally, I should mention some related work of Heitmann [2].
He has shown, under fairly general hypotheses, that from
a (possibly infinite) family  $\{R_i\}$  of Noetherian domains one can build
a Noetherian domain  R  such that  $spec(R) \cong \bigcup_i spec(R_i)$ , with only the
minimal elements identified.  This work is an outgrowth of a question,
raised informally by Kaplansky and more recently by Hochster [4]:  If
P  and  Q  are  ht 2  primes in a Noetherian domain, does  P ∩ Q  contain
a  **ht**  1  prime?  Counter-examples were obtained independently by
McAdam/Ratliff [9] and Heitmann.

§5.  Appendix:  The Maximal Spectrum.  Let  m-spec(R)  be the set of

maximal ideals of  R , with the relative Zariski topology.  It is

known [3] that the spaces of the form  m-spec(R) ,  R  a commutative

ring, are precisely the quasi-compact  $T_1$  spaces.  If  R  is Noetherian,

we see that  m-spec(R)  must be a Noetherian topological space, and

(*)  for each  x  there is a bound on lengths of chains of irreduc-

ible closed sets containing  x  (by the Krull height theorem).  Call a

$T_1$  Noetherian space satisfying  (*)  a  special  space.

Conjecture.  Every special space is homeomorphic to  m-spec(R)  for

some Noetherian ring  R .

     This conjecture has an equivalent and more intuitive formulation

in terms of the partially ordered set  j-spec(R)  of j-primes, that

is, primes that are intersections of maximal ideals.  Call a poset  X

special provided  X  satisfies (1), (2), (3) and (7) above, and, in

addition, each nonmaximal member of  X  has infinitely many covers.

It is easy to see that  j-spec(R)  is special, and the conjecture

above may be restated:  Every special poset is order-isomorphic to

the j-spectrum of a Noetherian ring.  (See [12] for the details.)

     One reason such a conjecture is not as reckless as it sounds is

that there is no principal ideal theorem for j-primes.  To see this,

let  R  be a countable 1-dimensional local domain.  Then the j-spectrum

of  R[X]  looks like this:

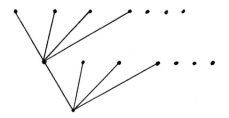

Obviously, condition (6) fails in this poset.

Here is some concrete evidence to support the conjecture:

Theorem [12].  Every special space with only countably many closed sets
is homeomorphic to the maximal ideal space of a countable Noetherian
ring.  Every countable special poset is order-isomorphic to the j-
spectrum of a countable Noetherian ring.

Acknowledgment.  I would like to thank Ray Heitmann and Sylvia
Wiegand for their indispensible help in preparing this paper.

## REFERENCES

1. Eisenbud, D. and E. G. Evans, Jr. "Every algebraic set in n-space is the intersection of n hypersurfaces," Invent. Math. 19(1973), 107-112.

2. Heitmann, R. C. "Prime ideal posets in Noetherian rings," Rocky Mtn. J. of Math., to appear.

3. Hochster, M. "Prime ideal structure in commutative rings," Trans. Amer. Math. Soc. 142(1969), 43-60.

4. _____. Topics in the Homology Theory of Modules over Commutative Rings, Amer. Math. Soc. Regional Conference Series 24(1975).

5. Kaplansky, I. Commutative Rings, Allyn & Bacon, Boston, 1970.

6. Krull, W. "Primidealketten in allgemeinen Ringbereichen," S.-B. Heidelberg Akad. Wiss. 7(1928).

7. Lewis, W. J. and J. Ohm. "The ordering of spec(R)," Canad J. Math., to appear.

8. McAdam, S. "Saturated chains in Noetherian rings," Indiana Univ. Math. J. 23(1974), 719-728.

9. _____. "Intersections of height 2 primes," J. of Algebra, to appear.

10. Nagata, M. "On the chain problem for prime ideals," Nagoya Math. J. 10(1956), 51-64.

11. _____. Local Rings, Interscience, New York, 1962.

12. Wiegand, S. and R. Wiegand. "The maximal ideal space of a Noetherian ring," J. Pure and Applied Algebra, to appear.

13. Zariski, O. and P. Samuel. Commutative Algebra, Vol. II, Van Nostrand, 1960.

# DENSE RINGS OF LINEAR TRANSFORMATIONS

J. M. Zelmanowitz

University of California, Santa Barbara

§1.  **Introduction.**  The purpose of this article is to present a partial
exposé of extensions of the Jacobson density theorem and the concomitant
theory of primitive rings beginning with the original contribution of
Jacobson [7].

**Jacobson Density Theorem.**  For a ring  R , the following conditions are
equivalent.

   (1)  R  has a faithful simple right module.

   (2)  There exists a vector space  $_\Delta V$  over a division ring  $\Delta$
        such that  $R \subseteq \mathrm{End}(_\Delta V)$  and given any elements  $v_1,\ldots,v_n$
        in  V  independent over  $\Delta$  and any elements  $u_1,\ldots,u_n$  in
        V , there exists  r  in  R' with  $v_i r = u_i$  for all  i .

An equivalent rephrasal of (2) which will be important in the
sequel is:

   (3)  $R \subseteq \mathrm{End}(_\Delta V)$  for some vector space  $_\Delta V$ , and given any finite

dimensional subspace  U  of  $_\Delta V$  and  $\tau$  in  $\text{End}(_\Delta V)$ , there
exists  r  in  R  with  $r|_U = \tau|_U$ .

A ring with a faithful simple right module is called (right) _primitive_; and Condition (2) or (3) defines  R  to be a **dense ring** of linear transformations in  $\text{End}(_\Delta V)$ .  We will also consider the following characterization of primitive rings [9, p.33].  _A right primitive ring is either isomorphic to_ $\Delta_t$ , _the_ t × t _matrix ring over a division ring_ $\Delta$ , _for some_ t (this is the case when  $_\Delta V$  in (2) above can be chosen finite dimensional); _or else for each positive integer_ t _there is a subring_ $R^{[t]}$ _of_ R _and a homomorphism of_ $R^{[t]}$ _onto_ $\Delta_t$ .

An important special class of primitive rings was treated by Jacobson in [8] as follows.

**Theorem 1.**  For a ring  R  the following conditions are equivalent.

(1)   R  has a faithful minimal one-sided ideal.

(2)   R  is isomorphic to a dense ring of linear transformations
      in some  $\text{End}(V_\Delta)$  which contains a nonzero linear transforma-
      tion of finite rank.

(3)   There exists a nondegenerate Morita context  $(R, _R V_\Delta, _\Delta V'_R, \Delta)$
      with  $\Delta$  a division ring and with  $_R V$  faithful.

For the definition of and basic properties of Morita contexts, see [1] or [2].

Associated to the class of primitive rings there is, of course, the
Jacobson radical, $J(R)$ , which may be defined as the intersection of all
ideals I of R such that R/I is primitive.

In what follows we will discuss some of the generalizations and
extensions of these theorems that have appeared to date.

§2. **Prime Nonsingular Rings.** The utility of the theory of primitive
rings in general and of the Jacobson radical in particular has been
amply demonstrated in the literature. So it was not long before an
effort was made to extend the density theorem to embrace a wider class
of prime rings, primitive rings obviously being prime.

This program was initiated by R. E. Johnson with another purpose in
mind; he was attempting to produce what are now called the Goldie
Theorems [4]. In a sequence of papers, most notably in [10], Johnson
essentially showed (in present terminology) that a prime nonsingular
(i.e., having right singular ideal equal to zero) ring with a uniform
right ideal satisfied a certain weakened density criterion. This
material was reworked and expanded by Amitsur in [1] with the aid of the
machinery of Morita contexts, and earlier by Koh and Mewborn in [11].
The central theorem can be stated as follows. It represents a generalization of Theorem 1.

**Theorem 2.** The following conditions on a ring R are equivalent:

(1) R is a prime ring with a nonsingular uniform right ideal.

(2)   R can be embedded in some $\text{End}(V_\Delta)$ for $\Delta$ a division ring in such a way that given any $\tau$ in $\text{End}(V_\Delta)$ and any finite dimensional subspace $U$ of $V_\Delta$, there exists $r$ and $s$ in $R$ with $\tau r = s$, $rV \subseteq U$, and $r|_U$ an automorphism of $U_\Delta$.

(3)   There exists a nondegenerate Morita context $(R, {}_R M_D, {}_D M'_R, D)$ where $D$ is a right Ore domain, $M_D$ is a torsion-free module, and ${}_R M$ is faithful.

Moreover, when (3) holds, setting $\Delta$ equal to the right quotient division ring of $D$ and $V = M \otimes_D \Delta$, then (2) holds for $V_\Delta$ in the following strengthened form. Given a finite dimensional subspace $U$ of $V_\Delta$, there exists a basis $u_1, \ldots, u_n$ of $U$ contained in $M$ and there exist $r$ and $s$ in $R$ and $0 \neq a$ in $D$ such that $\tau r = s$ and $ru_i = u_i a$ for each $i$.

This class of rings is broad enough to include orders in simple Artinian rings; in fact, they arise in (2) of the theorem precisely when $V_\Delta$ is finite dimensional. This theorem does not, however, include the Jacobson density theorem. When specialized to primitive rings it yields precisely the class of primitive rings with minimal right ideals.

§3.   **The Semiprime Case.** The Jacobson density theorem is readily generalized to rings having faithful semisimple modules [9, p.127]. Such rings are semiprime, and are dense subrings of direct products of full linear rings (the definition of density being extended to this situation in the obvious manner).

For semiprime nonsingular rings the theorem of Johnson and Amitsur described in §2 and other results in [1] were given in [15]. A definition is needed at this point. Call a module $M_R$ **semiprime** provided that for each $0 \neq m$ in M there exists f in $\text{Hom}_R(M,R)$ with $m(f(m)) \neq 0$ .

**Theorem 3.** The following conditions on a ring R are equivalent:

    (1)  R is a semiprime nonsingular ring containing a faithful direct sum of uniform right ideals.

    (2)  R has a faithful nonsingular semiprime module which is a direct sum of uniform modules.

    (3)  There exist division rings $\Delta_i$ and right $\Delta_i$-vector spaces $V_i$ , i in I , with R isomorphic to a subring of $\prod_{i \in I} \text{End}((V_i)_{\Delta_i})$ which is dense in the sense of Theorem 2(2).

    (4)  There exists a nondegenerate Morita context $(R, {}_R M_D, {}_D M'_R, D)$ where $D = \oplus \sum_{i \in I} D_i$ , $M = \oplus \sum_{i \in I} M_i$ , each $M_i$ is a torsion-free module over the right Ore domain $D_i$ , and ${}_R M$ is faithful.

One interesting byproduct of this endeavor arises by specializing condition (1) to the semisimple case. This gives an extension of Theorem 1 to the semiprime case [15, Corollary 4.4] which reads as follows.

**Theorem 4.** For a ring R the following conditions are equivalent:

(1)   R   has a faithful semisimple right ideal.

(2)   There exist division rings   $\Delta_i$   and vector spaces   $(V_i)_{\Delta_i}$ ,
      i  in  I , such that   R   can be embedded as a dense subring
      of   $\prod_{i \in I}$  $\text{End}((V_i)_{\Delta_i})$ .

§4.   **Rings with Critically Compressible Modules**.  An attempt to produce
a theorem simultaneously embracing primitive rings and prime Goldie rings
was made by Koh and Mewborn in [13].  A definitive summation of their wor
appears in [14].  We summarize the main result.

A ring  R  is called **right weakly transitive** provided there exists
a vector space   $_\Delta V$ , a right order  D  in  $\Delta$ , and a uniform  D-R  submodule  M  of  V   with       $\Delta M = V$ , such that if  $m_1, \ldots, m_t$  in  M
are linearly independent over  D  and  $n_1, \ldots, n_t$  are in  M , there exist
r  in  R  and  $0 \neq a$  in  D  with  $m_i r = a n_i$  for each  i .

A right ideal  I  of  R  is **almost maximal** provided

(a)   R/I  is uniform;

(b)   if  $J_1$  and  $J_2$  are right ideals properly containing  I ,
      then so does  $N(I) \cap J_1 \cap J_2$  where  $N(I) = \{r$  in  R $\mid rI \subseteq I\}$

(c)   if  $J \supsetneq I$  and  $aI \subseteq J$  then  $(J:a) \supsetneq I$  where  $(J:a) = \{r$  in
      ar  in  J} .

In [12] it was proved that **a ring**  R  **is right weakly transitive if
and only if it has an almost maximal right ideal**  I  **such that**  R/I  **is
faithful**.  Furthermore, if one defines  W(R)  to be the intersection of

all ideals  I  such that  R/I  is right weakly transitive, then  W(R)  is
a radical for rings in the usual sense,  $W(R) \subseteq J(R)$ , and  W(R)
contains every nil one-sided ideal.  (This can be found in [13].)

   We will call a nonzero module  $M_R$  **critically compressible**
provided that for any nonzero submodule  N  of  $M_R$ ,

   (a)  $\text{Hom}_R(M,N)$  contains a monomorphism, while

   (b)  $\text{Hom}_R(M,M/N)$  contains no monomorphisms.

A module satisfying only condition (a) is called  **compressible**.  For a
compressible module  $M_R$  each of the following two conditions is equivalent
to  $M_R$  being critically compressible:

   (1)  $M_R$  is monoform (i.e., nonzero partial endomorphisms of  $M_R$
        are monomorphisms);

   (2)  $M_R$  is uniform and every nonzero endomorphism of  $M_R$  is a
        monomorphism.

   In [6] A.Heinicke improved on the work cited above by showing among
other interesting results that  R  **is a right weakly transitive ring if
and only if**  R  **has a faithful critically compressible right module.**
(Heinicke actually used a characterization of these as compressible
modules which are rational extensions of their nonzero submodules.)

   To simplify the rest of this presentation let us call  $(\Delta, {}_\Delta V_R, M_R)$
an  **R-lattice** if  V  is a  $\Delta$-R bimodule with  $\Delta$  a division ring,
$\Delta M = V$ , and  R  acts faithfully on  M  (so that we can assume
$R \subseteq \text{End}({}_\Delta V)$ ).

Theorem 5.  (Density Theorem).  The following conditions are equivalent
for a ring  $R$ :

(1)  $R$  has a faithful critically compressible right module.

(2)  There exists an R-lattice  $(\Delta, {}_\Delta V_R, M_R)$  such that given any
elements  $m_1, \ldots, m_t$  in  $V$  which are linearly independent
over  $\Delta$ , there exists  $0 \neq a$  in  $\Delta$  such that for any elements  $n_1, \ldots, n_t$  in  $M$  one can find  $r$  in  $R$  with
$an_i = m_i r$  in  $M$  for each  $i$ .

(3)  There exists an R-lattice  $(\Delta, {}_\Delta V_R, M_R)$  such that given any
$\tau$  in  $\text{End}({}_\Delta V)$  and any elements  $m_1, \ldots, m_t$  in  $M$  which are
linearly independent over  $\Delta$ , there exists  $r$  and  $s$  in
$R$  with  $m_i \tau r = m_i s$  and  $0 \neq m_i r$  in  $\Delta m_i$  for each  $i$ .

We remark that when  $M_R$  is critically compressible,  $D = \text{End}(M_R)$
is a right Ore domain with quotient division ring  $\Delta = \text{End}(\bar{M}_R)$ , where
$\bar{M}_R$  denotes the quasi-injective hull of  $M$ ; and the element  $a$  in  $\Delta$
in (2) above can be chosen to lie in  $D$ .

Proof.  We will present a sketch.  Full demonstrations of this and the
remaining results will be found in [16].

(1) implies (2):  This is not much more difficult than the corresponding part of the Jacobson density theorem.  As mentioned above
$D = \text{End}(M_R)$  is a right order in the division ring  $\Delta = \text{End}(\bar{M}_R)$ , and
we can regard  $R$  as a subring of  $\text{End}({}_\Delta \bar{M})$ .  Set  $V = \bar{M}$ .  Let
$m_1, \ldots, m_t$  in  $V$  be given as in (2).  Then by a well-known lemma for

quasi-injective modules, $A_i = \bigcap_{j \neq i} (0:m_j) \nsubseteq (0:m_i)$ for each $i$, where

$(0:m_i) = \{r \text{ in } R \mid m_i r = 0\}$. Then $N = \bigcap_{i=1}^{t} m_i A_i \cap M \neq 0$ and since

$M_R$ is compressible there exists $0 \neq a$ in $D$ with $aM \subseteq N$. Given

$n_1, \ldots, n_t$ in $M$, $an_i = m_i r_i$ for some $r_i$ in $A_i$. Setting

$r = \sum_{i=1}^{t} r_i$ gives the desired conclusion.

(2) implies (3): One proceeds by induction on $k$, the induction

hypothesis being that there exist $r'$ and $s'$ in $R$ with $m_i \tau r' = m_i s'$

for $i < k$ and with $0 \neq m_i r$ in $\Delta m_i$ for $i = 1, \ldots, t$. The initial

case $k = 1$ is just condition (2). Replacing $\tau$ by $\tau r' - s'$, one

may in fact assume that $m_i \tau = 0$ for $i = 1, \ldots, k - 1$. The cases

$m_k \tau$ in $\sum_{i=1}^{k} \Delta m_i$ and $m_k \tau$ not in $\sum_{i=1}^{k} \Delta m_i$ are treated separately by

several applications of (2) to produce elements $r$ and $s$ in $R$ with

$m_i s = 0$ for $i < k$ and $m_k \tau r = m_k s$.

(3) implies (1): It is quite straightforward to show that if (3)

holds, then every cyclic submodule of $M_R$ is compressible and monoform.

This completes the proof.


We call a ring $R$ (right) **weakly primitive** if it has a faithful

critically compressible module. Further $R$ is a **weakly dense** ring of

linear transformations in $\text{End}(_\Delta V)$ when (2) holds. Condition (3) of

this theorem provides a suitable definition of a "local order" in a ring

of linear transformations; this is because $\tau = sr^{-1}$ when restricted to the $\Delta$-subspace generated by $m_1, \ldots, m_t$ . We remark that in the course of proving "(2) implies (3)" it is possible to choose $0 \neq a$ in $\Delta$ with $m_i r = a m_i$ for each $i$ .

From condition (3) it is readily apparent that this class of rings includes right orders in simple Artinian rings. It is only slightly less obvious that the weakly primitive rings include the prime nonsingular rings of Theorem 2. In this connection we can provide some information which further elucidates the parallel relationship between Theorem 1 and Theorem 2.

<u>Theorem 6</u>. The following conditions are equivalent for a ring $R$ :

   (1)   $R$ is a prime ring with a nonsingular uniform right ideal.

   (2)   $R$ has a faithful critically compressible right ideal.

   (3)   $R$ is a weakly dense dense ring of linear transformations
         possessing a nonzero linear transformation of finite rank.

<u>Proof</u>. (1) implies (3): This is contained in Theorem 2.

   (3) implies (2): The weak density of $R$ forces $R$ to be a prime ring. Next choose $r$ in $R$ with the rank of $r$ minimal. The weak density condition allows one to conclude that the rank of $r$ equals $1$ . The proof of this implication is completed by showing that $rR$ is a critically compressible right ideal of $R$ .

   (2) implies (1): This follows from two facts. First, the annihilator of a compressible module is a prime ideal; and secondly, that a monoform right ideal of a prime ring is nonsingular.

The characterization of primitive rings described in the Introduction generalizes as follows. If R is a weakly primitive ring then either R is a right order in a matrix ring $\Delta_t$ over some division ring $\Delta$ , in which case R contains a subring isomorphic to $D_t$ for some right order D in $\Delta$ ; or else for each positive integer t there exists a subring $R^{[t]}$ of R and a homomorphism of $R^{[t]}$ onto $D_t^{[t]}$ for some right order $D^{[t]}$ of $\Delta$ .

The proof of this can be outlined as follows. Let $M_R$ be a faithful critically compressible module and suppose that $m_1,\ldots,m_t$ in M are linearly independent over $\Delta = \text{End}(\overline{M_R})$ . Let $D = \{\bar{d}$ in $\Delta \mid dm_j$ in $\sum\limits_{i=1}^{t} m_i R$ for all $j\}$ , and verify that D is a right order in $\Delta$ . Let $A_i = \bigcap\limits_{j\neq i} (0:m_j)$ and $N = \bigcap\limits_{i=1}^{t} m_i A_i$ . Set $E = \{d$ in $\Delta \mid dm_j$ in N for all $j\}$ . Then E is a nonzero right ideal of D and hence a right order in $\Delta$ . Define $W = \sum\limits_{i=1}^{t} Em_i$ , and show that an element of $\text{End}(_E W)$ is given by right multiplication by an element in R . Put $R^{[t]} = \{r$ in $R \mid Wr \subseteq W\}$ , a subring of R , and $A^{[t]} = \{r$ in $R \mid Wr = 0\}$ , an ideal of $R^{[t]}$ . Then $R^{[t]}/A^{[t]} \cong \text{End}(_E W) \cong E_t$ which is a right order in $\Delta_t$ . Finally, observe that if $_\Delta V$ is of dimension t , then $A^{[t]} = 0$ .

In the finite dimensional case this reduces to a characterization of right orders in simple Artinian rings which includes also the Faith-Utumi theorem [3].

<u>Corollary</u>.  The ring  R  is a right order in a simple Artinian ring if
and only if  R  has a faithful critically compressible module  $M_R$  such
that  $\dim(_\Delta \Delta M) < \infty$ , where  $\Delta = \text{End}(\bar{M}_R)$  (a division ring).

This corollary was also proved independently in [6].

Examples of critically compressible modules arise from the observation that a Noetherian compressible module is critically compressible.
More generally, in [5, p.11] it is proved that a compressible module with
Krull dimension is critically compressible.  An example of a weakly dense
ring which is neither primitive nor contains a transformation of finite
rank is afforded by the free ring in two variables over a commutative
integral domain.  This can be proved by modifying the argument in [9,
p.36] which shows that the free algebra in two generators over a field
is primitive.

§5.  <u>The Weak Radical, Some Questions</u>.  Define  W(R)  to be the intersection of all ideals  I  of  R  such that  R/I  is weakly primitive.
Then  W(R)  is called the <u>weak</u> <u>radical</u> of  R .  As was mentioned earlier
in §4, with a slightly different definition,  W(R)  is a radical for
rings in the usual sense, and contains every nil one-sided ideal.  One
other interesting property of  W(R)  is that <u>when</u>  R  <u>is a ring with</u>
<u>Krull dimension</u>  W(R)  <u>is nilpotent</u>.

The proof of this is straightforward.  By a standard argument  W(R)
equals the intersection of the annihilators of all critically compressible

modules.  As we noted earlier, these annihilators are prime ideals of

R  so  W(R)  contains the prime radical.  Now for any prime ideal  P

of  R ,  R/P  is a prime ring with Krull dimension, hence is right Goldie

[5, p.20].  Thus  R/P  is weakly primitive, and it follows from the

definition of  W(R)  that  W(R) ⊆ P .  So  W(R)  equals the prime radical,

which is nilpotent in a ring with Krull dimension [5, p.32].

We can give other descriptions of the weak radical analogous to the

various characterizations of the Jacobson radical.  There is no need to

go into this here.  But the most important characterization still eludes

us; namely, an analogue for the description of  J(R)  in terms of

quasi-regularity.  This is not the only embarrassment in this area.  Note that

W(R)  is defined in terms of right weak primitivity.  There is also a

radical  W'(R)  defined similarly for left weak primitivity.  We conclude

with these two open problems.

(1)  Is there a characterization of  W(R)  in terms of "weak quasi-
regularity"?

(2)  Does  W(R) = W'(R) ?

A decent affirmative solution to question (1) should settle (2).

REFERENCES

1. Amitsur, S. A. "Rings of quotients and Morita contexts," J. Algebra 17(1971), 273-298.

2. Bass, H. The Morita Theorems, Lecture Notes, University of Oregon, 1962.

3. Faith, C. and Y. Utumi. "On Noetherian prime rings," Trans. Amer. Math. Soc. 114(1965), 53-60.

4. Goldie, A. W. "Semiprime rings with maximum conditions," Proc. London Math. Soc. 10(1960), 201-220.

5. Gordon, R. and J. C. Robson. Krull Dimension, Amer. Math. Soc. Memoir No. 133, Providence, R. I., 1973.

6. Heinicke, A. "Some results in the theory of radicals of associative rings," Ph.D. Thesis, University of British Columbia, 1969.

7. Jacobson, N. "Structure theory of simple rings without finiteness assumptions," Trans. Amer. Math. Soc. 57(1945), 228-245.

8. _____. "On the theory of primitive rings," Ann. of Math. 48 (1947), 8-21.

9. _____. Structure of Rings, Amer. Math. Soc. Colloq. Publ. Vol. 37, Providence, R. I., 1964.

10. Johnson, R. E. "Representations of prime rings," Trans. Amer. Math. Soc. 74(1953), 351-357.

11. Koh, K. and A. C. Mewborn. "Prime rings with maximal annihilator and maximal complement right ideals," Proc. Amer. Math. Soc. 16 (1965), 1073-1076.

12. _____.  "A class of prime rings," <u>Canad</u>. <u>Math</u>.

Bull. 9(1966), 63-72.

13. _____.  "The weak radical of a ring," <u>Proc</u>.

<u>Amer</u>. <u>Math</u>. <u>Soc</u>. 18(1967), 554-559.

14. Mewborn, A. C.  "Quasi-simple modules and weak transitivity," in

<u>Ring Theory</u>, R. Gordon, Ed., Academic Press, New York, 1972, 241-249.

15. Zelmanowitz, J.  "Semiprime modules with maximum conditions," <u>J</u>.

<u>Algebra</u> 25(1973), 554-574.

16. _____.  "Weakly primitive rings," to appear.